OZONE-FORMING POTENTIAL OF REFORMULATED GASOLINE ·

COMMITTEE ON OZONE-FORMING POTENTIAL OF
REFORMULATED GASOLINE

BOARD ON ENVIRONMENTAL STUDIES AND TOXICOLOGY

BOARD ON ATMOSPHERIC SCIENCES AND CLIMATE

COMMISSION ON GEOSCIENCES, ENVIRONMENT, AND RESOURCES

NATIONAL RESEARCH COUNCIL

D1300807

NATIONAL ACADEMY PRESS
WASHINGTON, D.C.

NATIONAL ACADEMY PRESS   2101 Constitution Ave., N.W.   Washington, D.C. 20418

NOTICE: The project that is the subject of this report was approved by the Governing Board of the National Research Council, whose members are drawn from the councils of the National Academy of Sciences, the National Academy of Engineering, and the Institute of Medicine. The members of the committee responsible for the report were chosen for their special competencies and with regard for appropriate balance.

The National Academy of Sciences is a private, nonprofit, self-perpetuating society of distinguished scholars engaged in scientific and engineering research, dedicated to the furtherance of science and technology and to their use for the general welfare. Upon the authority of the charter granted to it by the Congress in 1863, the Academy has a mandate that requires it to advise the federal government on scientific and technical matters. Dr. Bruce Alberts is president of the National Academy of Sciences.

The National Academy of Engineering was established in 1964, under the charter of the National Academy of Sciences, as a parallel organization of outstanding engineers. It is autonomous in its administration and in the selection of its members, sharing with the National Academy of Sciences the responsibility for advising the federal government. The National Academy of Engineering also sponsors engineering programs aimed at meeting national needs, encourages education and research, and recognizes the superior achievements of engineers. Dr. William A. Wulf is president of the National Academy of Engineering.

The Institute of Medicine was established in 1970 by the National Academy of Sciences to secure the services of eminent members of appropriate professions in the examination of policy matters pertaining to the health of the public. The Institute acts under the responsibility given to the National Academy of Sciences by its congressional charter to be an adviser to the federal government and, upon its own initiative, to identify issues of medical care, research, and education. Dr. Kenneth I. Shine is president of the Institute of Medicine.

The National Research Council was organized by the National Academy of Sciences in 1916 to associate the broad community of science and technology with the Academy's purposes of furthering knowledge and advising the federal government. Functioning in accordance with general policies determined by the Academy, the Council has become the principal operating agency of both the National Academy of Sciences and the National Academy of Engineering in providing services to the government, the public, and the scientific and engineering communities. The Council is administered jointly by both Academies and the Institute of Medicine. Dr. Bruce M. Alberts and Dr. William A. Wulf are chairman and vice chairman, respectively, of the National Research Council.

This project was supported by Contract No. 68D60069 between the National Academy of Sciences and the Environmental Protection Agency. Any opinions, findings, conclusions, or recommendations expressed in this publication are those of the author(s) and do not necessarily reflect the view of the organizations or agencies that provided support for this project.

International Standard Book Number 0-309-06445-7

Additional copies of this report are available from:

National Academy Press
2101 Constitution Ave., NW
Box 285
Washington, DC 20055
800-624-6242
202-334-3313 (in the Washington metropolitan area)
http://www.nap.edu

iv

OTHER REPORTS OF THE
BOARD ON ENVIRONMENTAL STUDIES AND TOXICOLOGY

Research Priorities for Airborne Particulate Matter: I. Immediate Priorities and a
  Long-Range Research Portfolio (1998)
The National Research Council's Committee on Toxicology: The First 50 Years
  (1997)
Toxicologic Assessment of the Army's Zinc Cadmium Sulfide Dispersion Tests
  (1997)
Carcinogens and Anticarcinogens in the Human Diet: A Comparison of Naturally
  Occurring and Synthetic Substances (1996)
Upstream: Salmon and Society in the Pacific Northwest (1996)
Science and the Endangered Species Act (1995)
Wetlands: Characteristics and Boundaries (1995)
Biologic Markers [Urinary Toxicology (1995), Immunotoxicology (1992),
  Environmental Neurotoxicology (1992), Pulmonary Toxicology (1989),
  Reproductive Toxicology (1989)]
Review of EPA's Environmental Monitoring and Assessment Program (three
  reports, 1994-1995)
Science and Judgment in Risk Assessment (1994)
Ranking Hazardous Waste Sites for Remedial Action (1994)
Pesticides in the Diets of Infants and Children (1993)
Issues in Risk Assessment (1993)
Setting Priorities for Land Conservation (1993)
Protecting Visibility in National Parks and Wilderness Areas (1993)
Dolphins and the Tuna Industry (1992)
Hazardous Materials on the Public Lands (1992)
Science and the National Parks (1992)
Animals as Sentinels of Environmental Health Hazards (1991)
Assessment of the U.S. Outer Continental Shelf Environmental Studies Program,
  Volumes I-IV (1991-1993)
Human Exposure Assessment for Airborne Pollutants (1991)
Monitoring Human Tissues for Toxic Substances (1991)
Rethinking the Ozone Problem in Urban and Regional Air Pollution (1991)
Decline of the Sea Turtles (1990)
Tracking Toxic Substances at Industrial Facilities (1990)

Copies of these reports may be ordered from
the National Academy Press
(800) 624-6242 or (202) 334-3313
www.nap.edu

Preface

THE CLEAN AIR ACT requires the use of reformulated gasoline (RFG) in specific areas of the United States with substantial ozone-pollution problems in an effort to make emissions from light-duty motor vehicles (automobiles and small trucks) less ozone forming and less toxic. That act requires RFG to have a minimum oxygen content of 2% (by weight) to promote more-extensive combustion of ozone-forming pollutants. Methyl tertiary-butyl ether (MTBE) and ethanol are two of the most widely used oxygenates that are blended into RFG to attain the oxygen requirement.

The U.S. Environmental Protection Agency (EPA) has established emission performance standards for RFG blends based on the *mass* of emissions of volatile organic compounds. Because ethanol-blended gasoline has a higher volatility than other blends and thus results in increased evaporation of organic compounds, it is difficult for such blends to meet the RFG standards unless the ethanol is blended with special low-volatility gasoline, which is more expensive and not readily available in many markets.

Proponents for the increased use of ethanol in RFG believe that the effects of the increased volatility of ethanol blends could be offset by the benefits that might be achieved through a reduction in ozone-forming potential. It is believed that emissions from the use of ethanol blends of RFG are less reactive in the atmosphere. However, EPA has no established method to assess RFG blends on the basis of ozone-forming potential.

Some members of Congress have been urging EPA to consider certifying RFG blends based on atmospheric reactivity or ozone-forming potential of the resulting emissions—not on just the mass of emissions as is done now. At the urging of Senator Lugar and others, EPA arranged for this study with the National Research Council (NRC). The Committee on Ozone-Forming Potential for Reformulated Gasoline was formed in 1997 by the NRC in response to the request from EPA.

The committee was asked whether the existing body of scientific and technical information is sufficient to permit a robust evaluation and comparison of the emissions from motor vehicles using different reformulated gasolines based on their ozone-forming potentials and to assess the concomitant impact of that approach on air-quality benefits of the use of oxygenates within the RFG program. As part of its charge, the committee was asked to consider (1) the technical soundness of various approaches for evaluating and comparing the relative ozone-forming potentials of RFG blends, (2) technical aspects of various air-quality issues related to RFG assessment, and (3) the sensitivity of evaluations of the relative ozone-forming potentials to factors related to fuel properties and the variability of vehicle technologies and driving patterns.

It is important to note that the committee was not asked to consider scientific issues beyond air quality, such as the relative health risks related to human exposure to various blends of RFG and their resulting emissions. Also, the committee was not asked to address the political, economic, and legal ramifications of changing the way that RFG certification is carried out.

The committee was generously assisted by many people, including those who presented valuable information and documents during the committee's public sessions: Charles Freed, Susan Willis, and Christine Brunner, U.S. EPA; Dean Simmeroth and Lawrence Larsen, California Air Resources Board (CARB) staff; Dennis Lawler, Illinois EPA; Michael Ward, of Swidler and Berlin; Gary Whitten, Systems Application International; Alan Dunker, General Motors; Cal Hodge, Oxygenated Fuels Association Technical Committee; Barry McNutt, U.S. Department of Energy; William Carter, University of California at Riverside; Robert Harley, University of California at Berkeley; Howard Feldman, American Petroleum Institute; Charles Schleyer, Mobil. Special thanks are due to Patricia McElroy and Robert Beaver of the University of California at Riverside, and Kevin Cleary of CARB staff who provided valuable assistance in data analysis. Also, Robert Dinneen, Renewable Fuels Association; Stephen Cadle, Coordinating Research Council; and Jose Gomez,

CARB staff, provided very useful information at the committee's request.

This report has been reviewed in draft form by individuals chosen for their diverse perspectives and technical expertise in accordance with procedures by the NRC's Report Review Committee. The purpose of this independent review is to provide candid and critical comments that assist the NRC in making the published report as sound as possible and to ensure that the report meets institutional standards for objectivity, evidence, and responsiveness to the study charge. The content of the final report is the responsibility of the NRC and the study committee and not the responsibility of the reviewers. The review comments and draft manuscript remain confidential to protect the integrity of the deliberative process. We wish to thank the following individuals for their participation in the review of this report: David Allen, University of Texas at Austin; Bart Croes, CARB staff; Richard Derwent, Meteorological Office, Berkshire, U.K.; Alan Dunker, General Motors; Thomas Graedel, Yale University; Robert Harley, University of California at Berkeley; Harvey Jeffries, University of North Carolina at Chapel Hill; Douglas Lawson, National Renewable Energy Laboratory; Thomas Peterson, University of Arizona; F. Sherwood Rowland, University of California at Irvine; Marc Ross, University of Michigan; Charles Schleyer, Mobil; Lance Waller, Emory University; and Gary Whitten, Systems Application International.

The individuals listed above have provided many constructive comments and suggestions. It must be emphasized, however, that responsibility for the final content of this report rests entirely with the authoring committee and the NRC.

The committee was ably assisted by NRC staff, especially Raymond Wassel, James Reisa, James Zucchetto, Laurie Geller, K. John Holmes, Robert Crossgrove, Ruth Danoff, Tracie Holby, and others.

Finally, I would like to express my thanks to the members of the committee for their diligent work. This report reflects the committee's consensus response to its charge.

William Chameides,
Chair, Committee on Ozone-Forming
Potential for Reformulated Gasoline

Contents

Ozone-Forming Potential of Reformulated Gasoline

Executive Summary

THE FEDERAL REFORMULATED GASOLINE (RFG) Program was mandated by the Clean Air Act Amendments of 1990 (Public Law 101-549) to help mitigate near-ground ozone pollution, a principal component of "smog," in the United States. In the lower atmosphere, ozone is produced by chemical reactions involving nitrogen oxides (NO_x), a great variety of volatile organic compounds (VOCs), and carbon monoxide (CO) in the presence of sunlight. All three types of ozone-precursor compounds are emitted by gasoline-fueled motor vehicles, so the control of motor vehicle emissions has been a major emphasis of the nation's effort over several decades to address the problem of ozone pollution.

The RFG program attempts to lower motor-vehicle emissions through re-engineering gasoline blends. For example, the Clean Air Act mandates a specified minimum oxygen content in RFG blends to help reduce emissions of ozone precursors from gasoline-fueled motor vehicles and to reduce the need for some toxic compounds, such as benzene, in the fuel. By itself, conventional gasoline has no oxygen content. Therefore, oxygen-containing chemical additives, called oxygenates, are blended into the fuel.

Implementation of the RFG program has involved controversy about how to determine which RFG formulations meet the various requirements of the program and which do not. The use of oxygenates in RFG is perhaps the most contentious aspect. Methyl tertiary-butyl ether (MTBE) and ethanol are two of the oxygenates most commonly used to meet the RFG program's oxygen requirement. One aspect of the controversy involves the release of toxic compounds into the environment; for example, a phase-out of MTBE has already been mandated in California

1

because of concern about environmental risks associated with MTBE leakage into drinking water. The other aspect of the controversy, which is the focus of this report, relates to the ozone pollution problem. MTBE and ethanol can affect the amounts and types of ozone precursor compounds emitted from tailpipes of motor vehicles as well as from the evaporation of unburned fuel. Questions persist over which oxygenate is preferable in terms of air-quality impact. This report addresses the potential impact of oxygenates in RFG on the ozone-forming potential of emissions from motor vehicles.

How should regulatory agencies determine if one RFG blend using a particular oxygenate is preferable to another? In attempting to mitigate ozone pollution, the U.S. Environmental Protection Agency (EPA) currently addresses such questions by estimating the *mass* of VOC emissions resulting from the use of an individual RFG blend. If the estimated mass of emissions exceeds a specified amount, that fuel blend is disallowed. However, a different method for assessing RFG blends has been proposed. Although certain fuel blends, such as those using ethanol, might result in greater amounts of emissions in terms of mass (because of the volatility of ethanol), it is argued that those emissions have a lower ozone-forming potential compared with emissions from other fuel blends. Therefore, the argument goes, EPA's assessment of RFG blends should be based not only upon *mass* of emissions, but also upon their *reactivity* (i.e., ozone-forming potential).

To help assess the scientific underpinning for this question, EPA asked the National Research Council to study it independently. In response, the Research Council formed the Committee on Ozone-Forming Potential of Reformulated Gasoline, which has prepared this report. The committee was charged to assess the utility and scientific rigor of evaluating the ozone-forming potential of the emissions resulting from RFG use (i.e., an approach that takes into account not only the total mass of emissions, but also the potential of the emissions to produce ozone). The committee was not charged or constituted to address the design or implementation of possible new regulations based on the ozone-forming potential of RFG blends. In addition, the committee was not charged or constituted to address relevant, but separate, issues about domestic sources versus foreign sources of fuel, relative energy and cost implications for the production of different RFG blends, relative health and global environmental impacts, or the use of renewable versus non-renewable fuels.

In approaching the task addressed by the committee, it is useful to note the context that has led to the RFG program and, thus, to the need for this study. Efforts to reduce ozone pollution in the United States have clearly had a positive impact on our nation's air quality. After accounting for the effects of meteorological fluctuations, data from EPA's Aerometric Information Retrieval System indicate that peak ozone concentrations in 41 metropolitan areas in the United States decreased by about 10% overall from 1986 to 1997 despite growing fuel usage. Nevertheless, ozone pollution remains a problem; in 1997, about 48 million people lived in areas of the United States that were classified as ozone "non-attainment" areas, and promulgation of the new 8-hr National Ambient Air Quality Standard (NAAQS) of 0.08 parts per million (ppm) for ozone is projected to triple the number of counties in non-attainment and to result in extensive non-attainment in rural areas of the eastern United States. The persistence of ozone pollution has sparked a need for innovative approaches to mitigation, and the RFG program is one such attempt.

An assessment of the ozone-forming potential of emissions from motor vehicles fueled by RFG requires information on the types and amounts of emissions from the vehicles. Gasoline-fueled vehicles emit VOCs, NO_x, and CO. VOCs from engine exhaust include many different compounds, some of which are not present in the original fuel but are created in combustion. VOCs can also evaporate from a vehicle's fuel system, and are thus independent of combustion. Each type of VOC can react differently in the atmosphere and thus affect the overall ozone-forming potential of vehicular emissions. NO_x and CO emissions are generated during combustion and occur only in the exhaust.

In addition to what and how much is emitted, evaluating the ozone-forming potential of various RFG blends involves assessing how reactive the emitted pollutants might be in the chemical processes that form ozone in the lower atmosphere. If the effect of RFG on air quality is large, then the difference between two blends of RFG might be readily discernible. On the other hand, if RFG has a very small effect on air quality, it is likely to be very difficult to identify which of two RFG blends is preferable in terms of air-quality impacts, let alone to quantify these effects reliably.

With both its charge and the context in mind, the committee undertook a review and analysis of relevant data and literature and also considered written and oral statements from numerous experts from the academic, private, and public sectors. The major findings of these deliberations and analyses are summarized below.

1. OZONE-PRECURSOR EMISSIONS FROM GASOLINE-FUELED VEHICLES

Overall emissions of ozone precursors from gasoline-fueled motor vehicles have substantially decreased in recent decades, largely as a result of government mandates and industry's development and use of new emission controls on motor vehicles.

According to EPA estimates for 1997, emissions of VOCs from on-road gasoline-fueled motor vehicles contributed about 26% to the total inventory of VOC emissions from all sources. Correspondingly, on-road vehicles contributed 22% to the inventory for NO_x, and 56% for carbon monoxide (CO). These contributions are projected to continue to shrink in the coming years. If correct, this would imply that the potential impact of using RFG on near-ground ozone concentrations will decrease with time. In fact, air-quality models suggest that implementation of the RFG program reduces peak ozone concentrations by only a few percent. Even if the relative contribution of motor vehicles to the current inventory of ozone precursor emissions from all sources has been underestimated (which, historically, has often been the case), the reduction in peak ozone from the RFG program would still likely be less than 10% at most. Although long-term trends in peak ozone in the United States appear to be downward, it is not certain that any part of these trends can be significantly attributed to the use of RFG.

2. HIGH-EMITTING MOTOR VEHICLES

A sizable portion of the ozone-precursor emissions from gasoline-fueled motor vehicles appears to be associated with a relatively small number of high-emitting vehicles in the United States.

Emissions tests, tunnel studies, and remote-sensing of tailpipe exhaust suggest that a disproportionately large fraction of motor-vehicle exhaust emissions arise from a relatively small number of high-emitting vehicles. Many such vehicles have improperly functioning catalyst systems because of catalyst deterioration or improper control of the air-to-fuel ratio. In addition, tests performed during the operation of motor vehicles indicate that a substantial contribution of emissions occurs during a cold start (i.e., before the catalyst system reaches its operating temperature). The committee did not have sufficient information to assess whether vehicles with malfunctioning evaporative-control systems also are important contributors. The great majority of emissions testing of motor vehicles using RFG has been performed on normally functioning vehicles, and there is substantial uncertainty over how RFG affects emissions from high emitting vehicles. Therefore, it is difficult to quantify total motor-vehicle emissions for an entire motor-vehicle fleet and to assess the efficacy of the use of RFG for the full driving cycle.

3. THE USE OF REACTIVITY IN ASSESSING THE OZONE-FORMING POTENTIAL OF EMISSIONS

The use of reactivity in assessing the ozone-forming potential of VOC emissions has reached a substantial level of scientific rigor, largely as a result of research sparked by policy making in California over the past several decades.

Ozone chemistry involves many thousands of reactions and a similar number of compounds. Not only does ozone formation respond differently to different VOC compounds and different amounts of NO_x, it also responds differently in different locations or pollution episodes. Assessment of reactivity is most appropriate for VOC-limited areas (i.e., areas where ozone concentrations are more sensitive to changes in VOC concentrations than to changes in NO_x concentrations). It is likely that reactivity factors could be used in those areas to address non-attainment of the new 8-hr, 0.08-ppm NAAQS for ozone, in a manner similar to that used to address non-attainment of the current 1-hr, 0.12-ppm NAAQS. However, it should be noted that in NO_x-limited regions, reactivity—as it is currently used—is of limited value with respect to ozone mitigation. Little research has been undertaken on the derivation and application of NO_x reactivity.

4. RELATIVE REACTIVITY AS A MEANS OF COMPARING FUELS

The most robust reactivity measures for comparing emissions from different sources are the so-called relative-reactivity factors, but they are often uncertain and of limited utility for comparing similar RFG blends.

These factors are formed by taking the ratio of the reactivity of one compound or emission source to that of another, and thereby canceling out many of the uncertainties associated with the calculation of reactivities. Even so, relative-reactivity factors are typically subject to substantial uncertainty. Various studies suggest that the uncertainty in relative reactivity for emissions, such as those arising from motor vehicles, is generally about 15-30% (at the 95% level of statistical confidence). The major contributors to this uncertainty arise from the substantial variability and difficulty in characterizing how different vehicles respond to changes in fuel composition, the limited amount of test data available, and the limited knowledge of how well a vehicle fleet is characterized by the available data. Because the reactivity of emissions from motor vehicles using various RFG formulations tends to be quite similar and the emissions composition so variable, the reactivity approach is sometimes of limited utility.

5. REACTIVITY OF CO EMISSIONS

CO in exhaust emissions from motor vehicles contributes about 20% to the overall reactivity of motor-vehicle emissions.

The contribution of CO to ozone formation should be recognized in assessments of the effects of RFG. If adding an oxygenate to a gasoline significantly changes the amount of CO emitted by the motor vehicle fleet, this would affect ozone formation. Further, as VOC emissions from mobile sources continue to decrease in the future, CO emissions might become proportionately an even greater contributor to ozone formation. The committee did not conclude that the various RFG oxygenates affected CO emissions to such a degree that they substantially altered reactivity comparisons between RFG blends. However, it is important to note that there are substantial uncertainties in how fuel oxygen impacts CO emissions from high-emitters, as well as in the contribution of high-emitters to overall CO emissions.

6. OVERALL AIR-QUALITY BENEFIT OF RFG

Emissions tests, tunnel studies, and remote-sensing of tailpipe exhaust indicate that RFG usage can cause a decrease in both the exhaust and evaporative emissions from motor vehicles.

In addition to a minimum oxygen content, the RFG program requires gasoline blends to have a number of other characteristics that are intended to produce lower emissions. Major contributors to decreased emissions appear to be lowering the Reid Vapor Pressure (RVP)[1] of the fuel, which helps depress evaporative emissions of VOC, and lowering the concentration of sulfur in the fuel, which prevents poisoning of a vehicle's catalytic converter by sulfur. Overall, it is estimated that use of RFG can result in approximately a 20% reduction in the mass and total reactivity of VOC emissions from motor vehicles. In addition, such blends can lead to reductions in emissions of CO and some air toxics. Despite such emission reductions, however, the overall effect of the RFG program on ozone air quality is expected to be difficult to discern.

7. EFFECT OF OXYGENATES IN RFG

The use of commonly available oxygenates in RFG has little impact on improving ozone air quality and has some disadvantages.

Although there is some indication that oxygenates decrease the mass of VOC and CO exhaust emissions, as well as their combined reactivity, the decrease, if any, appears to be quite small. Moreover, some data suggest that oxygenates can lead to higher NO_x emissions, which are more important than VOC emissions in determining ambient ozone levels in some areas. Thus, the addition of commonly available oxygenates to RFG is likely to have little air-quality impact in terms of ozone reduction.

The most significant advantage of oxygenates in RFG appears to be a displacement of some toxics (e.g., benzene) from the RFG blend, which results in a decrease in toxic emissions. However, not all air toxics are decreased; for example, emissions of formaldehyde are not decreased and might even be increased by MTBE blends of RFG. Although ethanol blends of RFG might not increase formaldehyde emissions, they lead to increased emissions of acetaldehyde.

8. MTBE BLENDS VERSUS ETHANOL BLENDS—
EXHAUST EMISSIONS

The reactivity of the exhaust emissions from motor vehicles operating on

---

[1]RVP is the constrained vapor pressure of a fuel at 100°F.

ethanol-blended RFG appear to be lower—but not significantly lower—than the reactivity of the exhaust emissions from motor vehicles operating on MTBE-blended RFG.

Data from emission tests indicate that there is no statistically significant difference (at the 95% confidence level) between RFGs blended with MTBE or ethanol in the mass of VOC or NO_x exhaust emissions from motor vehicles. There is also no statistically significant difference between MTBE and ethanol blends in the reactivity of VOC exhaust emissions. No evidence supports the claim that reactivity-weighted VOC emissions from properly operating motor vehicles using RFG with ethanol would be significantly less than those from motor vehicles using RFG blended with MTBE, even if the ethanol-containing fuel had more oxygen than the MTBE-containing fuel. On the other hand, some data indicate that exhaust emissions of CO from motor vehicles using RFG blended with ethanol are somewhat lower than those of motor vehicles using an MTBE-blended RFG. As a result, a small reduction in the reactivity of the combined VOC and CO exhaust emissions from motor vehicles might result from the use of an ethanol-blended RFG over that of a MTBE-blended RFG.

9. MTBE BLENDS VERSUS ETHANOL BLENDS—EVAPORATIVE EMISSIONS

Both the mass and reactivity (mass of ozone per mile) of evaporative emissions from motor vehicles using ethanol-blended RFG were significantly higher than from motor vehicles using MTBE-blended RFG.

The higher evaporative emissions of the ethanol-blends were likely due, at least in part, to the fact that such blends had an RVP that is approximately 1 pound per square inch (psi) higher than the equivalent MTBE-blended fuel. Moreover, the increase in total reactivity of evaporative emissions from the ethanol-blended RFG far outweighed the small decrease in the reactivity of the exhaust emissions described in Finding 8. As a result, it appears that a net increase in the overall reactivity of motor-vehicle emissions (exhaust plus evaporative) would result from the use of ethanol-blended RFG (with an elevated RVP) instead of MTBE-blended RFG.

10. REID VAPOR PRESSURE OF ETHANOL-CONTAINING FUEL

On the basis of Finding 9 above, it appears likely that the use of an ethanol-containing RFG with an RVP that is 1 psi higher than other RFG blends would be detrimental to air quality in terms of ozone.

This conclusion is consistent with the California Air Resources Board's 1998 evaluation that led to its decision against allowing a 1 psi RVP-waiver for ethanol-containing fuels. However, it should be borne in mind that (1) the committee's conclusion is based on tests using properly functioning motor vehicles and, thus, might or might not have underestimated the benefits of using ethanol-blended RFG in high-emitting vehicles; and (2), as discussed earlier, the overall impact on ozone of allowing the use of ethanol-containing fuel would likely be quite small in any case.

11. USE OF REACTIVITY TO EVALUATE RFGS

The committee sees no compelling scientific reasons at this time to recommend that fuel certification under the RFG program be evaluated on the basis of the reactivity of the emission components.

Analyses of available data on emissions from the use of ethanol-blended RFG and MTBE-blended RFGs showed that mass-emissions differences between the two fuels varied on occasion from the differences found by using reactivity as a basis. However, in no case was the fundamental conclusion concerning the choice of one fuel over another on the basis of relative potential air-quality benefits altered by switching from a mass-emissions metric to a reactivity-weighted metric.

12. MODELS USED TO CHARACTERIZE EMISSIONS FROM RFG BLENDS

The models currently used to inform regulatory decision making—by quantifying emissions from motor vehicles that use RFG blends—are problematic.

The current models are based on regression equations developed from data obtained from a limited set of tests on a small sampling of motor vehicles. Although the Complex and Predictive models are distinct from models used to estimate the mobile source inventory, their capability of reflecting actual emissions needs to be improved. In some cases, algorithms used to develop the regression equations for the models ignore important parameters that can influence emissions. For example, the Complex Model, developed by EPA, does not account for temperature variations when calculating evaporative emissions. The Predictive Model, developed by the California Air Resources Board, excludes consideration of evaporative emissions. Another potential source of error in both models arises from their treatment of high-emitting vehicles. As noted above, a large portion of motor-vehicle emissions come from high-emitting vehicles. However, the emissions from these vehicles are likely to be quite variable and thus difficult to characterize through sampling a small subset of the total population.

13. OPPORTUNITY TO TRACK EFFECTS OF PHASE II RFG PROGRAM

The scheduled implementation of Phase II of the federal RFG program in 2000 offers a unique opportunity to track and document the impact of a new ozone-mitigation program. Plans should be made and implemented for an atmospheric measurements program to assess the impact of Phase II RFG on (1) emissions of ozone precursors from the on-road and non-road motor vehicle fleet, as well as ozone-forming potential of those emissions; and (2) the impact of these changes, if any, on ambient concentrations of ozone and its precursors.

1

Introduction

PHOTOCHEMICAL SMOG, and its concomitant high concentrations of ground-level ozone (O_3) and other noxious compounds, is caused by a complex series of chemical reactions involving the oxidation of volatile organic compounds (VOCs)[1] and carbon monoxide (CO) in the presence of nitrogen oxides (NO_x) and sunlight (Figure 1-1). As illustrated in Figure 1-2, the transportation sector is responsible for a large fraction of VOC, CO, and NO_x emissions in the United States. On-road gasoline-fueled motor vehicles are estimated to account for about 26% of the VOC emissions from all source categories, about 56% of the CO emissions, and about 22% of the NO_x emissions in 1997 (EPA 1998). As a result, motor vehicles have been a primary target for emission controls in the

---

[1]An organic compound is a compound containing carbon combined with atoms of other elements, commonly hydrogen, oxygen, and nitrogen. Simple carbon-containing compounds such as carbon monoxide (CO) and carbon dioxide (CO_2) are usually classified as inorganic compounds. A volatile organic compound (VOC) is an organic compound that exists as a gas under typical atmospheric conditions. A large number of acronyms are used to denote various categories of volatile organic compounds; a listing of some of the more common acronyms and their meanings is presented in Chapter 3. In this report, organic compounds in the gas phase are referred to as "VOCs" unless noted otherwise.

12

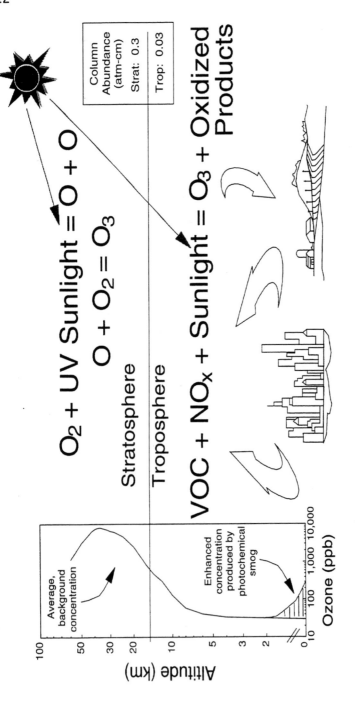

FIGURE 1-1 Atmospheric ozone (O_3). Left-handed graphic: average O_3 mixing ratio (in parts per billion by volume) as a function of altitude; center: photochemical processes responsible for producing O_3 in the stratosphere and troposphere; right-handed inset: column abundance of O_3 in atm-cm (i.e., the height of the O_3 column if it were compressed to 1 atmosphere pressure). Although only VOCs and NO_x are shown as ozone precursors, carbon monoxide (CO) can participate, in much the same way as VOCs, in the sequence of reactions leading to ozone formation. Source: Adapted from EPA 1998.

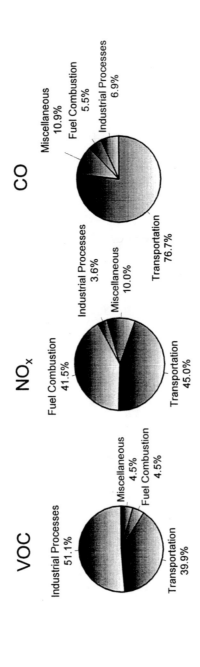

FIGURE 1-2 19,214,000 tons of VOC, 23,582,000 tons of NO$_x$, and 87,451,000 tons of CO emissions in 1997 in the United States by source category. Emissions from on-road vehicles are included in the transportation source category. Source: EPA 1998.

nation's strategy for mitigation of the ozone pollution problem.[2, 3] As part of this effort, the Clean Air Act Amendments of 1990 called for the development and use of reformulated gasoline (RFG)[4] in light-duty motor vehicles to reduce the ozone precursor emissions from those vehicles.

Through the Clean Air Act Amendments of 1990, Congress mandated that RFG contain at least 2.0% oxygen by weight to decrease the emissions of ozone-precursors and air toxics. To meet that requirement, RFG blends typically contain small amounts of additives referred to as oxygenates, which are organic compounds that contain some chemically bound oxygen. The use of these oxygenates in RFG has given rise to a complex and often contentious debate concerning the relative benefits of one oxygenated compound over another (e.g., methyl tertiary-butyl ether (MTBE) versus ethanol). Oxygenates can affect the amounts and types of ozone precursors emitted by motor vehicles in different and potentially offsetting ways. Is oxygenated gasoline preferable from an air-quality point of view over nonoxygenated gasoline? Is one oxygenated compound clearly preferable from an air-quality point of view over another? Should some oxygenated additives be allowed to be used in RFG whereas others should not be allowed? In the traditional approach to ozone mitigation in the United States, these questions are addressed in a straightforward and simple manner: the mass of precursor emissions from the use of various RFG blends in motor vehicles are assessed; and,

---

[2]EPA estimates that non-road gasoline-fueled motor vehicles account for about 9% of the VOC emissions from all source categories, about 19% of the CO emissions, and about 19% of the NO_x emissions in 1997 (EPA 1998). Because emission test data from non-road vehicles fueled by RFG were not available to the committee, consideration of such emissions were not included in this study.

[3]Uncertainties associated with mobile source emission estimates are discussed in Chapter 4.

[4]In this report, "RFG" is used in the most generic sense to refer to blends of gasoline that have been reformulated to change any of a multitude of gasoline-blend characteristics (e.g., blend content, oxygen content, sulfur content, and vapor pressure) and motor-vehicle-emissions characteristics. Such use of the term "RFG" should not be confused with the more narrow regulatory definitions of RFG as a gasoline blend that is compliant with the specific requirements of the federal RFG program or the California RFG program. Those regulatory definitions represent a subset of the range of possible reformulated gasolines.

if the mass of emissions from a blend exceeds some specified amount, it is disallowed. In this report, the committee assesses the utility and accuracy of an alternate approach based on evaluating the ozone-forming potential of ozone-precursor emissions. With such an approach, an RFG blend with a high rate of emissions based on mass might be considered acceptable if those emissions were of relatively low ozone-forming potential. It should be noted that this report is limited to the scientific and technical aspects of this issue; the possible design or implementation of regulations based on ozone-forming potential are not addressed.

THE OZONE-POLLUTION PROBLEM

For every billion atmospheric molecules, there are, on average, only about 300 ozone molecules. Despite this minute concentration, atmospheric ozone has a major impact on the environment. In the stratosphere, where about 90% of the atmosphere's ozone resides, it protects life from harmful ultraviolet radiation from the sun. In the upper part of the troposphere, ozone is critical to the oxidation process in the atmosphere by which a wide range of pollutants is removed from the air we breathe.

Ground-level ozone (i.e., at the lowest part of the troposphere) represents a small, but important, component of the total burden of ozone found in the troposphere. Although ground-level ozone concentrations in remote regions of the atmosphere are about 20 to 40 parts per billion by volume (ppb), those concentrations often exceed 100 ppb during episodes of increased air pollution and can rise above 200 ppb during severe episodes in urban areas.[5] When ozone concentrations reach increased levels, they can harm rather than sustain organisms.

---

[5]In this report, the abundance or concentration of atmospheric ozone will be expressed in terms of its volume mixing ratio; that is, the number of ozone molecules per unit volume of air divided by the total number of atmospheric molecules per unit volume of air. Thus, an ozone concentration of 1 ppb denotes an ozone abundance of 1 ozone molecule for each billion atmospheric molecules, and an ozone concentration of 1 ppm (i.e., parts per million by volume) denotes an ozone abundance of 1 ozone molecule for each million atmospheric molecules, and is equal to 1,000 ppb.

Adverse effects include impairment of lung function in humans and other animals, and damage to agricultural crops, forests, and other vegetation. It is this aspect of tropospheric ozone, as a ground-level pollutant, that forms the backdrop for this report.

HISTORY OF U.S. POLICIES TO MITIGATE OZONE POLLUTION

The phenomenon known as photochemical smog was first documented in the 1940s when air pollutants were found to be causing damage to vegetable crops grown in the Los Angeles area (Middleton et al. 1950). Soon after, Haagen-Smit and others showed that ozone was the primary oxidant in photochemical smog causing crop damage and that it was produced by photochemical reactions requiring the participation of two types of precursors: VOCs and NO_x (Haagen-Smit et al. 1951, 1953; Haagen-Smit 1952; Haagen-Smit and Fox 1954, 1955, 1956).

Subsequent observations revealed that photochemical smog and the concomitant high concentrations of ground-level ozone that accompanied the smog were not unique to Los Angeles but were common to most of the major metropolitan areas of the United States and elsewhere in the world. Those observations, along with medical and epidemiological studies documenting the adverse health effects of ozone at concentrations encountered during air-pollution episodes, provided the impetus for the promulgation of regulations designed to control or even eliminate the problem. Passage of the Clean Air Act of 1970 (Public Law 91-604) established National Ambient Air Quality Standards (NAAQS) for ozone and other criteria pollutants as well as a federally coordinated program to reach attainment of these standards within specific deadlines. With the persistence of the ozone problem, ever more stringent and costly air-pollution controls were promulgated by the Clean Air Act Amendments of 1977 (Public Law 95095) and 1990 (see Table 1-1).

In retrospect, it appears that the ozone mitigation policies our nation has embarked upon over the past 3 decades have had a positive impact. On average, peak ozone concentrations in urban areas of the United States appear to be on a downward trend (Figure 1-3) and the problem would undoubtedly be considerably more severe if controls had not been implemented (see e.g., Harley et al. 1997). The U.S. Environmental Protection Agency (EPA) reports that ozone concentrations decreased for the 1-hr and 8-hr averaging times (shown in Figure 1-3)

TABLE 1-1   Milestones in Ozone Pollution and Its Control in the United States

Year	Milestone	Notes
1840s	Ozone molecule discovered	Schoenbein 1840
1850s	Ozone presence in atmosphere documented	Schoenbein 1854
1874	Ozone shown to be toxic to animals	Andrews 1874
1940s	Photochemical smog found to be causing crop damage	Middleton 1950
1950s	Ozone found to be major oxidant in photochemical smog VOC's and NO_x shown to be ozone photochemical precursors	Haagen-Smit 1952
1961	Basic science of ozone pollution documented in monograph	Leighton 1961
1970	Clean Air Act of 1970 (CAA-70) establishes national program for the mitigation of ozone pollution in the United States. Sets 1975 as deadline for attainment of NAAQS	
1975	CAA-70 attainment deadline not met	
1977	Clean Air Act Amendments of 1977 (CAAA-77) establishes 1982 and 1987 as new deadlines for attainment	
1987	CAAA-77 attainment deadlines still not met	
1990	Clean Air Act Amendments of 1990 establishes new attainment deadlines extending into the 21st century and authorizes implementation of a reformulated gasoline program	
1997	New 8-hr, 80-ppb NAAQS for ozone promulgated	EPA 1997a

on average by about 1% per year from 1986 to 1997 (EPA 1998). On the other hand, the problem remains far from solved. In 1997, about 48 million people lived in 77 counties where ozone concentrations exceeded the second daily maximum 0.12-ppm, 1-hr NAAQS for ozone (EPA 1998). Of the 29 urban areas required by the Clean Air Act Amendments of 1990 to submit State Implementation Plans, 27 were unable to submit plans that showed attainment by the mandated date of 1998. Moreover, the promulgation of a new 8-hr, 80-ppb NAAQS for ozone in 1997 is expected to approximately triple the number of non-attainment counties

Concentration, ppm

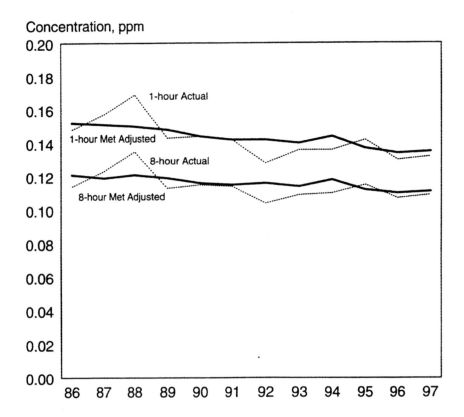

FIGURE 1-3 Comparison of actual (dotted lines) and meteorologically adjusted (solid lines) ozone trends in 1-hr and 8-hr 99th percentile ozone concentrations for the period of 1986-1997 across 41 metropolitan areas. Source: Adapted from EPA 1998.

and lead to widespread non-attainment in rural as well as urban areas of the eastern United States (Wolff 1996; Chameides et al. 1997). With the persistence of the ozone-pollution problem comes the need to develop new and innovative approaches to lowering ozone-precursor emissions. The federal RFG program is but one example of a new approach that is being promulgated to address this need.

In the formulation and testing of various blends of RFG, it became apparent that these blends could affect motor-vehicle emissions in various and subtle ways (AQIRP 1990; OTA 1990). In addition to affecting the total mass of VOC emissions, different RFGs could have different effects on the amounts of NO_x and CO emitted by motor vehicles. They could also affect the relative amounts of evaporative and exhaust emissions from motor vehicles and thus the chemical composition of the VOCs emitted by these vehicles. Controversy arose over whether the nation's traditional approach to assessing emissions, based on the mass of VOC emitted, was adequate to assess and regulate various RFG blends. With the use of ethanol as an oxygenated additive, such regulation proved to be especially contentious (e.g., EPA 1994). When compared with typical RFG blends using MTBE, blends using ethanol tend to have more evaporative VOC emissions but, it was argued, with a lower net ozone-forming potential. Accordingly, Senator Richard G. Lugar suggested that EPA establish a procedure to certify ethanol blends of RFG as equivalent to nonethanol blends based on ozone-forming potential (see Appendix B, letter from Senator Richard G. Lugar dated October 17, 1995). EPA has not done so because it was unsure that there was an appropriate method for making such an assessment. Instead, EPA has commissioned this report to address the scientific and technical bases for such an assessment.

CHARGE TO NATIONAL RESEARCH COUNCIL COMMITTEE

Does RFG with ethanol as the oxygenate result in vehicular emissions with a lower ozone-forming potential than a blend with MTBE? Can a metric based on ozone-forming potential be reliably and robustly used to quantify the relative impacts of different RFG blends with different oxygenates on ozone pollution in the United States? As outlined above, these are the questions that motivated the formation of the National Research Council Committee on Ozone-Forming Potential of Reformulated Gasoline and this report. More specifically, the committee was given the task to assess whether the existing body of scientific and technical information is sufficient to permit a robust evaluation and comparison of the emissions from motor vehicles using different RFG blends based on their relative ozone-forming potentials; and the concomitant impact on air-quality benefits of the RFG program. The committee was asked to focus its assessment on the use of oxygenates in RFG, with specific attention to RFG blends using MTBE and ethanol.

The committee was asked to address the following specific issues:

1.   Assessment of the technical soundness of various approaches

for evaluating and comparing the relative ozone-forming potentials of RFGs. Two prominent methods for assessing relative ozone impacts include relative reactivity factors and grid-based airshed models. Determine whether there is sound scientific basis for the use of reactivity factors, models, and/or any other approach(es) for evaluating the ozone-forming potential of RFGs in a nationwide program.

2.   Assessment of technical aspects of various air-quality issues when evaluating the relative ozone-forming potentials of RFGs. Air-quality issues to be considered include assessment of the ozone-forming potentials of RFGs for both peak (1-hr) and average (8-hr) ozone levels, inclusion or exclusion of CO as an ozone precursor, and consideration of changes in NO_x emissions and the corresponding impact on ozone levels resulting from the use of different levels and/or types of oxygenates or other fuel composition changes.

3.   Assessment of the sensitivity of evaluations of the relative ozone-forming potentials of RFGs to factors related to fuel properties and the variability of vehicle technologies and driving patterns. Factors to be considered include assessment of effects of the fuel blending method (i.e., splash blending versus match blending), "distillation impact" and/or the "commingling effect," variability in fuel composition, engine operating conditions as they pertain to emissions, and changes in the exhaust-to-evaporative emissions ratio.[6]

The committee was asked to identify any gaps in the existing scientific and technical information, recommend how such gaps might be

---

[6]"Splash blending" refers to a method of oxygenating gasoline by adding an oxygenate to the gasoline blend stock without any systematic control over the resulting Reid Vapor Pressure (RVP) of the RFG. (RVP is the constrained vapor pressure of the fuel at 100 degrees Fahrenheit.) When ethanol is splash blended into gasoline, the RVP of the finished blend could increase by 1 pound per square inch (psi) or more above the applicable RVP limit. "Match blending" refers to the preparation of an RFG blend with systematic control over the resulting RFG such that the finished blend meets the RVP standard for the appropriate RFG blend. "Distillation impact" refers to the possible effect of oxygenates on the volatility of RFG blends at temperatures greater than 100°F, which can occur in a vehicle's fuel tank. "Commingling effect' refers to an increase in the resulting vapor pressure when an ethanol-blended RFG is mixed with a non-ethanol blended RFG in a vehicle's fuel tank. The increase in vapor pressure of the mixture is beyond that of either of the separate blends.

filled, and identify the types of emission data that would be needed to continuously evaluate the ozone-forming potential of emissions from vehicles using RFG.

It is important to note that the committee was not asked to consider other issues related to the choice and use of various blends of RFG. Therefore, the committee has not addressed issues such as balance of trade, energy and cost requirements for fuel production, domestic sources of fuel versus foreign sources, human health and global environmental impacts, and use of renewable fuels versus nonrenewable fuels. In addition, it should be noted that this report is limited to the scientific and technical aspects of this issue; the possible design or implementation of regulations based on ozone-forming potential are not within the scope of this study.

REPORT STRUCTURE

In response to its charge, the committee's report addresses (1) how the ozone-forming potential of emissions from light-duty motor vehicles might be affected by the use of RFG blends with and without various types and concentrations of oxygenates; and (2) the extent to which current scientific and technical understanding and information are adequate to quantify these effects robustly. Although the focus is on the impacts of RFGs on ground-level ozone concentrations, RFG and the oxygenates added to these gasolines can also have impacts on other air-quality issues (e.g., toxics, carbon monoxide, and particulate matter); these other impacts are mentioned when they are relevant or potentially significant.

To provide a technical foundation for the assessment, the report provides overviews of the photochemistry of ozone, the concept of atmospheric reactivity and ozone-forming potential, motor vehicles as a source of ozone precursors, and RFGs in Chapters 2, 3, 4, and 5, respectively. In Chapter 6, the report assesses the likely magnitude of the air-quality benefits of the federal and California RFG programs (in total) based on observations. Chapter 6 also outlines the characteristics of a measurements program that could more robustly quantify the air-quality benefits and the changes in the ozone-forming potential of vehicular emissions arising from Phase II of the federal RFG program. The committee's assessment of RFG's overall impact on ozone serves as a calibration point for the discussion in Chapter 7, which focuses on eight specific RFG blends to illustrate the methodology of, as well as the advantages and problems associated with, using ozone-forming potential to evaluate the relative impacts of these blends. Appendix A contains biographical information on the committee. Appendix B contains a letter from Senator Richard G. Lugar suggesting that EPA establish a procedure to certify ethanol blends of RFG as equivalent to methanol blends based on ozone-forming potential. Appendix C presents the equation set for EPA's Complex Model of Phase II of the federal RFG program, and Appendix D presents motor-vehicle-emissions data evaluated by the committee.

2

Ozone Photochemistry

MITIGATION OF THE ozone-pollution problem is complicated by the fact that ozone (O_3) is a secondary pollutant; that is, it is not emitted directly into the atmosphere, but is produced by photochemical reactions involving primary pollutants and modulated by meteorological conditions. The problem is further confounded by the complex nature of the photochemical mechanism responsible for producing ozone and the intricate array of precursors that can participate in this photochemical mechanism. These complexities are briefly reviewed in this chapter.

VOC LIMITATION VS NO_x LIMITATION

As noted in Chapter 1, ozone is formed by chemical reactions involving volatile organic compounds (VOCs) and carbon monoxide (CO) in the presence of nitrogen oxides (NO_x) and sunlight. One might expect, therefore, that the severity of ozone pollution in a given region can be reduced by lowering the emissions of VOCs, CO, NO_x, or any combination thereof. However, mitigation of ozone pollution is not so straightforward. It turns out that the rate of ozone formation is a complex and variable function of the concentrations of VOC and NO_x as well as meteorological conditions. As a result, establishing the relative benefits of VOC and NO_x emissions controls can be a difficult and challenging task. The source of the complexity can be elucidated through an examination of

Figure 2-1, which is a schematic of the photochemical smog mechanism. Ozone production occurs as a result of a series of reactions initiated by the oxidation of VOCs or CO by the hydroxyl radical (OH). For example,

$$RH + OH \rightarrow R + H_2O \qquad\qquad (2\text{-}1)$$
$$R + O_2 + M \rightarrow RO_2 + M \qquad\qquad (2\text{-}2)$$
$$RO_2 + NO \rightarrow RO + NO_2 \qquad\qquad (2\text{-}3)$$
$$RO + O_2 \rightarrow HO_2 + \text{carbonyl} \qquad\qquad (2\text{-}4)$$
$$HO_2 + NO \rightarrow OH + NO_2 \qquad\qquad (2\text{-}5)$$
$$2x\,(NO_2 + h\nu \rightarrow NO + O) \qquad\qquad (2\text{-}6)$$
$$2x\,(O + O_2 + M \rightarrow O_3 + M) \qquad\qquad (2\text{-}7)$$
$$RH + 4O_2 + 2\,h\nu \rightarrow 2O_3 + \text{Carbonyl} + H_2O \qquad\qquad \text{NET}$$

where *RH* represents a generic hydrocarbon (or VOC), *R* is a hydrocarbon

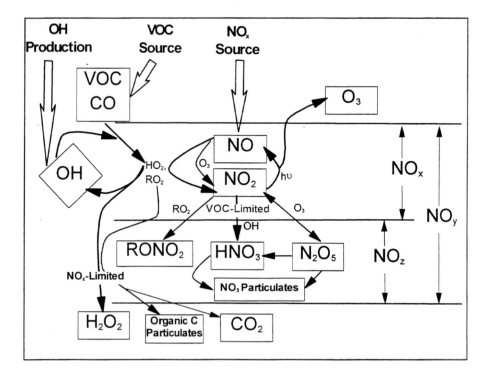

FIGURE 2-1 Schematic of the photochemical pathways leading to the production of ozone and the termination steps that dominate under NO_x-limited and VOC-limited regimes.

radical (e.g., CH_3CH_2 for RH = ethane), M is a nonreactive, energy-absorbing third body (N_2, O_2), and $h\nu$ represents energy from solar radiation (it is the product of Planck's constant h, and the frequency, ν, of the electromagnetic wave of solar radiation). Of note in this sequence is that VOCs are consumed, whereas both OH/HO$_2$ and NO$_x$ act as catalysts. Moreover, the by-product labeled "carbonyl" is itself a VOC and can, in general, react and produce additional ozone molecules. It is important to note that although OH is removed in Reaction 2-1, it is regenerated in Reaction 2-5.

Termination of the above ozone-generating cycle occurs when the catalysts are removed. Two important paths are

$$HO_2 + HO_2 + M \rightarrow H_2O_2 + O_2 + M, \text{ or} \qquad (2\text{-}8)$$
$$OH + NO_2 + M \rightarrow HNO_3 + M \qquad (2\text{-}9)$$

In general, the rate of ozone production can be limited by either VOCs or NO$_x$. The existence of these two opposing regimes, often schematically represented in a so-called EKMA (Empirical Kinetic Modeling Approach) diagram (Figure 2-2), can be mechanistically understood in terms of the relative sources of OH and NO$_x$ (Kleinman 1994, in press). When the rate of OH production is greater than the rate of production of NO$_x$, termination of the reaction chain that produces ozone is dominated by Reaction 2-8 (see Figure 2-1). Under these conditions, NO$_x$ is in short supply; as a result, the rate of ozone production is NO$_x$-limited (i.e., ozone is most effectively reduced by lowering NO$_x$). Therefore, ozone concentrations are most effectively reduced by lowering NO$_x$ emissions, and subsequent concentrations of NO$_x$, instead of lowering emissions of VOCs. When the rate of OH production is less than the rate of production of NO$_x$, on the other hand, termination of the ozone-forming chain proceeds predominately via Reaction 2-9 (see Figure 2-1), NO$_x$ is relatively abundant, and ozone production is VOC-limited (i.e., ozone is most effectively reduced by lowering VOCs). Because this region is characterized by rapid loss of OH via Reaction 2-9, it is also referred to as being the radical-limited regime. Finally, between these two extremes (i.e., the NO$_x$- and VOC-limited regions) lies a transitional region, sometimes referred to as the ridge in an EKMA diagram. In this transitional region, ozone is about equally sensitive to VOCs and NO$_x$, but, compared within its sensitivity to VOCs in the VOC-limited region and its sensitivity to NO$_x$ in the NO$_x$-limited region, ozone is relatively insensitive to both.

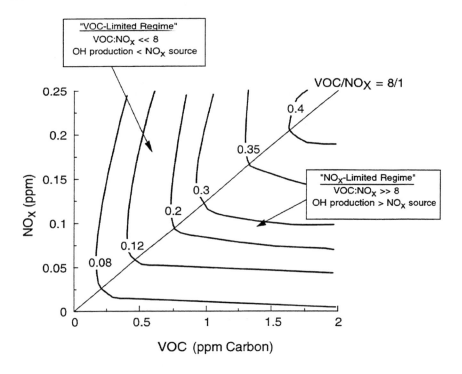

FIGURE 2-2 Typical EKMA (Empirical Kinetic Modeling Approach) diagram showing contours (or isopleths) of 1-hr maximum ozone concentrations (in parts per million by volume (ppm)) calculated as a function of initial VOC and NO_x concentrations and the regions of the diagram that are characterized by VOC limitation or NO_x limitation. "OH production" refers to the rate of OH photochemical production and "NO_x source" refers to the rate at which NO_x is emitted into the boundary layer.

A further complication arises from the fact that VOC and NO_x limitation is not uniquely defined by location or emissions. Instead, it is a chemical characteristic of an air parcel that varies dynamically with transport, dispersion, dilution, and photochemical aging. For example, consider the results of a series of photochemical box model calculations illustrated in Figure 2-3. In each calculation, a boundary-layer air parcel was assumed to have initial VOC and NO_x concentrations at 0800 hr and then allowed to react over the course of a single day while mixing with relatively clean air from aloft at varying rates. For simplicity, processes such as surface deposition and horizontal dispersion are not included. Although these simulations greatly simplify the photochemical smog

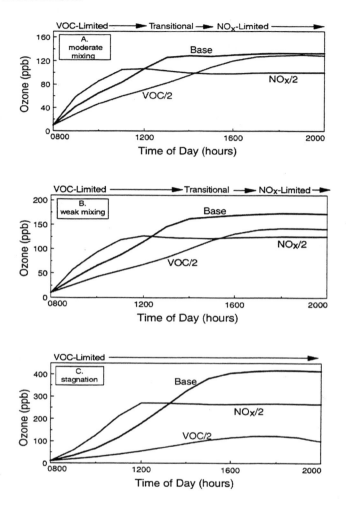

FIGURE 2-3 Model-calculated evolution of ozone as a function of time in an air parcel for various initial urban-like mixtures of VOCs and NO_x at 0800 hr under summertime conditions and three rates of vertical mixing and free tropospheric entrainment. For each mixing rate, simulations for three initial VOC and NO_x concentrations are presented: "Base" with initial VOC = 0.15 ppm and NO_x = 1.5 ppm; "VOC/2" with initial VOC = 0.75 and NO_x = 0.15 ppm; and "$NO_x/2$" with initial NO_x = 0.075 and VOC = 1.5 ppm. Note the characteristic tendency for the system to evolve from VOC limitation to NO_x limitation with time and for the point of transition to be delayed as mixing decreases. Also note that varying the initial conditions of either precursor in EKMA implicitly changes the emissions after 0800 hr by the same percentage.

phenomenon, they nevertheless capture much of the essence of the relationship between ozone and its precursors and are, therefore, useful to illustrate some key points.

In the first example (Figure 2-3A), a moderate amount of vertical mixing during a typical summer day is assumed. For these conditions and the high initial concentrations of VOCs and NO_x adopted for the Base case, the model predicts a rapid rise in ozone reaching a peak of about 130 ppb around mid-afternoon—an ozone variation that is characteristic of many moderate urban air-pollution episodes in the United States. (If the effects of dispersion and surface deposition are included, the peak concentration would have been somewhat depressed and the decay following the peak more pronounced.)

A key feature of the results illustrated in Figure 2-3A is the varying response of ozone to assumed decreases in the initial concentrations and emissions of VOCs and NO_x. Because of the nature of urban VOC and NO_x emissions, air parcels exposed to these emissions are usually initially within the VOC-limited regime. Thus, in Figure 2.3A, halving the initial VOC concentration is much more effective in reducing ozone than halving NO_x during the first ~5 hr of this particular simulation. In fact, during the first few hours of the simulation an "NO_x disbenefit" appears, that is, an increase in ozone results from a decrease in NO_x. This effect is caused by the conversion of more NO to NO_2, and by an increase in the fraction of OH radicals which react with VOCs (and thereby leading to RO_2 and HO_2 radicals, which convert NO to NO_2) compared with reaction with NO_2. For these conditions, a decrease in NO_x leads to more OH, more oxidation of VOCs (e.g., via Reaction 2-1), and thus an ozone increase.

Because NO_x is processed and removed rapidly, the NO_x disbenefit tends to be fairly short-lived. Moreover, as NO_x concentrations continue to fall, the air parcel begins to move from VOC limitation to the transitional region and often reaches NO_x limitation within many areas of the country. For the conditions adopted in the simulation illustrated in Figure 2-3A, ozone is more effectively reduced by halving NO_x than by halving VOCs after about 1400 hr. Another important feature of the calculations illustrated in Figure 2-3A, which is also characteristic of the photochemical smog system in general, is that the peak ozone concentration is reached when the air parcel is in the transitional region between VOC limitation and NO_x limitation. The formation of organic nitrates (including peroxyacetyl nitrate (PAN)) also affects ozone formation by removing NO_x from the system which would otherwise lead to ozone formations. Depending on the temperature, PAN formation can lead to a temporary reduction in ozone formation.

Other processes can further complicate and confound the relationship between ozone and its precursors. One of these is vertical mixing. As illustrated in Figure 2-4, vertical mixing has a direct impact on ozone concentrations: in the early morning hours it tends to contribute positively to ozone accumulation by bringing ozone-rich air from aloft into the boundary (or surface) layer, but in late morning and afternoon it tends to depress ozone by diluting surface air now laden with newly formed ozone with air from aloft. As a result, as the amount of mixing decreases and stagnation sets in, the severity of air-pollution episodes is exacerbated. That is illustrated in Figures 2-3B and 2-3C, in which higher peak ozone concentrations are generated as less vertical mixing and more stagnation occur. However, vertical mixing has another indirect, but still very important, effect on ozone. In addition to depressing peak ozone, vertical mixing also tends to depress NO_x concentrations in the polluted boundary layer by diluting it with cleaner air from aloft. For this reason, stagnation tends to slow the rate of transition from VOC limitation to NO_x limitation. If vertical mixing is extremely weak (i.e., conditions assumed for Figure 2-3C), the sun might set before NO_x is sufficiently processed to allow the parcel to make the transition from VOC limitation. Thus, the efficacy of VOC and NO_x controls is, in general, critically dependent upon the meteorological as well as the chemical conditions that prevail during any given episode.

The distribution of NO_x emissions can also affect where and if air parcels within a given airshed make the transition from VOC limitation. Like stagnation, the presence of dispersed NO_x sources in a large metropolitan area or megalopolis can lead to high NO_x concentrations throughout an area, fostering continuous VOC limitation.

REACTION PATHWAYS OF ETHANOL AND METHYL TERTIARY-BUTYL ETHER

For both methyl tertiary-butyl ether (MTBE) and ethanol, the important atmospheric loss processes are by reaction with the OH radical. Reaction of ethanol with the OH radical leads to the formation of acetaldehyde (CH_3CHO) in 100% or close to 100% yield (Atkinson 1994; Atkinson in

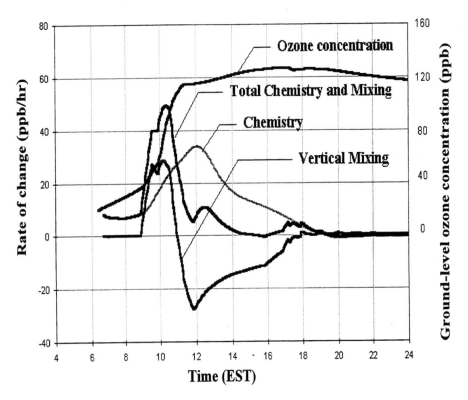

FIGURE 2-4 The relative contributions of vertical mixing and chemical production processes to the ground-level ozone concentration as a function of time during the day based on a one-dimensional model. No deposition processes are included in this simulation. The left scale is for the rate of change in ppb/hr and the right is for the ozone concentration in ppb. Source: Adapted from Zhang and Rao (In Press).

press), with the major reaction pathway (~90% of the overall OH radical reaction) proceeding by

$$OH + CH_3CH_2OH \rightarrow H_2O + CH_3\dot{C}HOH \qquad (2\text{-}10)$$
$$CH_3\dot{C}HOH + O_2 \rightarrow CH_3CHO + HO_2 \qquad (2\text{-}11)$$

For MTBE, the products of the OH radical reaction in the presence of NO are *tert*-butyl formate ($(CH_3)_3COCHO$), ~75%; formaldehyde, ~45%; methyl acetate ($CH_3C(O)OCH_3$), ~15%; and acetone, ~3% (see Atkinson 1994, and references therein). *tert*-Butyl formate reacts only slowly with the OH radical, with a half-life due to gas-phase reaction with the OH radical of around 11 days for a 24-hr average OH radical concentration of 1×10^6 molecules•cm^{-3}. Wet and dry deposition of *tert*-butyl formate may also be important.

SUMMARY

To the extent that these calculations can be generalized to represent the evolution of a plume as it advects from an urban center (or other concentrated sources of anthropogenic emissions of VOC and NO$_x$) to suburban and then rural areas (with time essentially representing distance from an urban core), they suggest that: (1) in isolated large urban cores and similar source regions, ozone concentrations during severe air-pollution episodes are most effectively reduced by reductions in VOC emissions and might even increase as a result of NO$_x$-emission controls; (2) ozone concentrations in rural areas and over large regional expanses are most effectively reduced by reductions in NO$_x$ emissions from the pollution sources that affect that area or region (e.g., upwind urban sources and important local sources); and (3) the highest ozone concentrations during an episode generally occur in locations somewhat removed from the major precursor source areas (i.e., suburban areas) and tend to occur when the chemistry of the system is in a transitional stage between VOC limitation and NO$_x$ limitation. Where the peak ozone concentration will occur during any given episode and whether it will occur when the chemistry is, in fact, transitional or is VOC-limited are determined by myriad factors including the meteorological conditions and distributions as well as intensities of emissions.

The fact that ozone formation can vary from VOC limitation to NO$_x$ limitation is highly germane to the topic of this report. As discussed in more detail in Chapter 3, ozone-forming potential has historically been used to characterize the ability of VOCs to produce ozone. Thus the relevance of using existing methods to assess the ozone-forming potential of various reformulated gasoline blends will be largely limited to those areas and episodes characterized by VOC limitation (or at least transitional chemistry).

A further complication in assessing the efficacy of emission controls for VOC and NO$_x$ arises from the fact that VOCs comprise a rich and varied assortment of compounds. Two principal VOC categories are those that arise from anthropogenic sources and those that arise from natural or biogenic sources (e.g., isoprene from trees and other vegetation).

Natural VOCs can participate in the photochemical reactions that produce ozone, but they cannot, in principle, be directly controlled like those from anthropogenic sources. In regions where natural VOCs represent a significant fraction of the total reactive VOCs, NO_x controls might be needed to reduce ozone substantially even if the oxidant chemistry is VOC-limited.

Moreover, the compounds that make up the general category of anthropogenic VOCs can be quite varied with widely different chemical characteristics and reactivities that lead to different rates of ozone formation. Thus, ton-for-ton, the reduction in the emissions of one VOC might lead to more or less reduction in ozone than the reduction of another VOC. The concept of ozone-forming potential, discussed in the next chapter, attempts to account for the differing chemical characteristics of VOCs as they relate to ozone photochemical production.

3

The Concept of Ozone-Forming
Potential and Its Quantification

IN ANY GIVEN AIRSHED, it is common to find hundreds of different VOC species, each with its own unique chemistry. In the simplest approach to ozone mitigation based on VOC controls, emission reductions are implemented on a mass basis without any regard to the unique chemistry of each of the VOCs. The principle behind ozone-forming potential or reactivity[1] is the notion that, in addition to the amount of a specific VOC species emitted into a given airshed, the difference in the chemistry of each of the VOCs needs to be considered when assessing the impact of those species on ozone formation.

The utility of the concept of ozone-forming potential can be illustrated through a comparison of the impacts on ozone concentrations in an urban airshed of two ubiquitous VOC species: ethane and propene.

---

[1]Because ozone-forming potential of a given VOC is dependent upon its propensity to react in the atmosphere, the term "reactivity" is often used to denote a species' ozone-forming potential. As discussed later in this chapter, terms such as kinetic reactivity (KR) and mechanistic reactivity (MR) are used to define specific processes that contribute to a species ozone-forming potential, whereas terms such as maximum incremental reactivity (MIR) are used to specify the method for calculating a species' ozone-forming potential.

33

If one were to increase the total mass of VOC emissions in a city, such as Los Angeles, by 20% through additional emissions of ethane, ozone levels would increase slightly. However, if the same amount of propene were added instead, there would be a large increase in ozone. Why the big difference between the two, given that both are rather simple hydrocarbons? The primary cause of the difference is the differing rates at which these two species react in the atmosphere. Ethane has an atmospheric lifetime of weeks. Little of the ethane emitted in an urban area reacts within that area before it is transported away. Its contribution to ozone formation within the urban area is therefore very small. Propene, on the other hand, has a lifetime of hours. Most of it will typically react near its source and thus be able to contribute to the photochemical production of ozone in the area in which it is emitted (or immediately downwind). A secondary, but smaller, cause for the differing impacts of the two species is the different number of ozone molecules formed in that environment for each molecule of ethane and propene that reacts. Differences in ozone productivity arising from the first effect are often expressed in terms of the kinetic reactivity (KR), and differences from the second are expressed in terms of the mechanistic reactivity (MR).

REGULATORY APPLICATION OF
VOC OZONE-FORMING POTENTIAL

There is, in fact, a significant historical precedent for accounting for VOC reactivity in U.S. regulatory policy, albeit to a limited extent (see Dimitriades 1996, for a history of VOC regulation in the United States). During the early years of ozone mitigation, it was recognized that there were some organics, for example ethane, that did not contribute significantly to smog formation on urban scales, whereas others, such as propene, did. Thus, two categories of organic gases were defined for regulatory purposes: unreactive and reactive[2] (see Table 3-1). However, the term

---

[2]Reactive VOCs are operationally designated as reactive organic gases (ROG). However, because hydrocarbons make up most of the organic gas emissions, this category is also referred to as reactive hydrocarbons (RHC). Moreover, because methane dominates the unreactive category, nonmethane hydrocarbons or NMHC is another term that is often used. These and other

TABLE 3-1 Acronyms and Names Used for Classifying Organic Compounds

Common Abbreviation	Full Name	Definition
VOC[a]	Volatile organic compound	Organic compounds that are found in the gas phase at ambient conditions. Might not include methane.
ROG	Reactive organic gas	Organic compounds that are assumed to be reactive at urban (and possibly regional) scales. Definitionally, taken as those organic compounds that are regulated because they lead to ozone formation. The term is predominantly used in California.
NMHC	Nonmethane hydrocarbon	All hydrocarbons except methane; sometimes used to denote ROG
NMOC	Nonmethane organic compound	Organic compounds other than methane
NMOG	Nonmethane organic gas	Organic gases other than methane
RHC	Reactive hydrocarbon	All reactive hydrocarbons; also used to denote ROG
THC	Total hydrocarbon	All hydrocarbons, sometimes used to denote VOC
OMHCE	Organic material hydrocarbon equivalent	Organic compound mass minus hetero-atom mass (i.e., carbon plus hydrogen mass only)
TOG	Total organic gas	Total gaseous organic compounds, including methane. Used interchangeably with VOC

[a]Unless noted otherwise, VOCs is the term used in this report to represent the general class of gaseous organic compounds.
Source: U.S. Environmental Protection Agency at http://www.epa.gov/docs/OCEPAterms.7

unreactive is a misnomer, because even compounds such as ethane and methane do react and contribute to tropospheric ozone formation,

terms are listed in Table 3-1. Unless noted otherwise, VOCs is the term used in this report to represent the general class of gaseous organic compounds.

though at much lower rates, on a per mass basis, than other compounds. Such low-reactivity compounds, particularly carbon monoxide (CO) and methane, do contribute to ozone formation, because emission rates of those compounds are very large. (The contribution of CO to ozone-forming potential is discussed further in Chapters 6 and 7.)

A complication in this two-category approach is deciding where to place the dividing line between unreactive and reactive VOCs. Some-what arbitrarily, that dividing line has been chosen to be at the level of reactivity of ethane. In the United States, but outside of California, species with reactivities equal to or less than that of ethane are placed in the unreactive category.

California has been a leading force in the application of reactivity assessment to ozone mitigation efforts. For example, California uses ozone-forming potential in its Low Emission Vehicles and Clean Fuels Program (LEV/CF) to adjust and regulate the amount of emissions from vehicles (CARB 1991). A fuel with higher VOC emissions, but a lower net reactivity than the reference fuel, is permitted in the program, thus providing an incentive to develop fuels with less-reactive emissions. (The current CARB program is limited, however, to exhaust emissions, and, as discussed in Chapter 4, evaporative emissions can be quite important.) The use of reactivity in California's regulatory air-quality programs has been a major catalyst for continuing research on ozone-forming potential and its application to policy-making. As the under-standing of how to define ozone-forming potential operationally has grown substantially in recent years, the use of ozone-forming potential to other regulatory issues (e.g., emissions from consumer products) is now under consideration.

OPERATIONAL DEFINITION OF OZONE-FORMING POTENTIAL USING REACTIVITY

The photochemical degradation of most VOC species is initiated by reaction with the OH radical (i.e., Reaction 2.1 in Chapter 2). Therefore, for most VOCs, the KR of a specific VOC is greater if its OH-radical reaction rate constant is greater. As seen in Table 3-2, these rate constants can vary by many orders of magnitude. A relatively simple type of reactivity scale, sometimes referred to as the OH-reactivity or the k_{OH} scale, expresses the relative contribution of VOCs in terms of their rates of reaction with OH (e.g., Darnall et al. 1976; Chameides et al. 1992).

TABLE 3-2   OH Rate Constants (k_{OH}) and Maximum Incremental Reactivity (MIR)[a] for Selected Compounds

Compound	$10^{12} \times k_{OH}$ [b] $(cm^3 \cdot molecule^{-1} s^{-1})$	MIR [c] O_3 formed/g VOC emitted[b]
Carbon monoxide	0.21	0.065
Methane	0.0062	0.016
Ethane	0.25	0.32
Propane	1.1	0.57
n-Butane	2.4	1.18
n-Octane	8.7	0.69
2,2,4-Trimethylpentane	3.6	1.34
Ethene	8.5	8.3
Propene	26	11.0
trans-2-Butene	64	13.2
Isoprene	101	9.3
∝-Pinene	54	3.9
Benzene	1.2	0.81
Toluene	6.0	5.1
m-Xylene	24	14.2
1,2,4-Trimethylbenzene	32	5.3
o-Cresol	42	2.5
Formaldehyde	9.4	6.6
Acetaldehyde	16	6.3
Acetone	0.22	0.49
2-Butanone	1.1	1.4
Methanol	0.94	0.65
Ethanol	3.3	1.7
Methyl tert-butyl ether	2.9	0.73
Ethyl tert-butyl ether	8.8	2.2
tert-Butyl formate	0.75	No value cited

[a]MIR combines kinetic and mechanistic reactivities of a specied compound for conditions that maximize the predicted reactivity of VOCs by making the reactive systems very NO_x rich.

[b]Rate constants at 298 K are taken from Atkinson (1994, 1997) and Le Calve et al. (1997)

[c]From Carter (1997), http://www.cert.ucr.edu/~carter/bycarter.htm. The MIR of the assumed urban mix used in the calculations was 4.06 g of O_3 per gram of VOC emitted.

This approach has some significant advantages. OH-rate constants for a large number of VOCs have already been well characterized by laboratory experiments, and many others can be estimated with a fair degree of reliability (e.g., Kwok and Atkinson 1995; Atkinson in press). Moreover, these constants are defined by the VOC species themselves and not the environment in which the VOCs are emitted (other than minor temperature dependencies). Thus, the OH reactivities for a wide range of VOC species can be readily calculated and compared. Combining these OH reactivities with data on the ambient concentrations of these VOCs provides a measure of the rate at which the various VOC species are oxidized and produce peroxy radicals (e.g., via Reaction 2-2 and Reaction 2-4 in Chapter 2), and thus provides a rough estimate of their relative potential roles in ozone-formation (Chameides et al. 1992).

There are, however, significant limitations to using the OH-reactivity scale to characterize the roles of VOCs: The method does not account for the potentially different yields of peroxy radicals formed from different VOCs, the different reactive pathways these peroxy radicals can take once they are produced, and the varying tendency of VOCs to enhance or inhibit radical levels, and thus influence the contribution of other VOC species to ozone formation. All of these factors can have a significant effect on the amount of ozone formed from the oxidation of a VOC (Carter and Atkinson 1989; Bowman and Seinfeld 1994; Carter 1994). For this reason, the OH-reactivity scale does not always correlate well with other measures of ozone-forming potential, particularly for the more rapidly reacting VOCs (e.g., Dodge 1984; Bergin et al. 1998a). For example, aromatics, which have strong NO_x sinks and radical sources in their mechanisms, can have relatively high reactivities under conditions with low ratios of VOC to NO_x, but negative values of reactivities when the VOC to NO_x ratio is sufficiently high, because NO_x, which (as NO_2) would otherwise photolyze to form ozone, is removed from the system.

MR is used to account for this second influence on the ozone-forming potential of VOCs (Carter and Atkinson 1989). In general, the variability in mechanistic reactivities is substantially less pronounced than that of kinetic reactivities, and thus the species-to-species variability of reactivity scales that combine KR and MR tend to follow the variability in KR but not exactly (see Table 3-2).

If KR is defined as the number of molecules of a specific VOC that react within a given airshed (by photolysis, reaction with the OH radical, reaction with NO_3 radical, or reaction with ozone) and MR is the number of ozone molecules that are formed for each VOC molecule in the system that reacts, the total number of ozone molecules formed from a given VOC molecule is equivalent to the product of the two quantities, that is,

$$\text{Ozone-forming potential} = KR \times MR \qquad (3\text{-}1)$$

This way of dissecting the ozone-forming potential of a compound, although remarkably simple, is also quite powerful and instructive. However, it also has its limitations. For example, neither KR nor MR is a property inherent in a compound. Instead, both are dependent upon the protocol established to calculate them (e.g., the type of environments in which the VOC exists and the length of time used to assess the amount of the VOC that reacts and the ozone that is formed). Thus, the use of the relationship expressed in Equation 3-1 requires an operational definition for quantifying reactivities.

QUANTIFYING OZONE-FORMING POTENTIAL USING REACTIVITY

If ozone-forming potential is to be used in ozone mitigation programs, it is necessary to develop an operational definition for ozone-forming potential, and a protocol for quantifying it. One such definition is the incremental reactivity (IR) proposed by Carter and Atkinson (1989) and Carter (1994).[3] IR is defined as the number of *additional* grams of ozone formed per gram of VOC compound *added* to a base mixture (the VOC compound could be present in the base mixture):

$$IR_i = \Delta[O_3]/\Delta[VOC_i] \qquad (3\text{-}2)$$

where IR_i is the incremental reactivity of species i; $\Delta[O_3]$ is the change in some ozone metric used to assess the impact of VOCs on air quality (e.g., the 1-hr peak or 8-hr averaged ozone concentration in an airshed) or the total human exposure to ozone above some threshold concentration); and $\Delta[VOC_i]$ represents a change in the emissions of species i

---

[3]Another scale, developed by Derwent and Hov (1979), is the photochemical ozone creation potential (POCP) scale. It is used to quantify the ozone-forming potential of VOC emissions. In general, the IR approach and the POCP approach produce qualitatively similar results.

(e.g., from an RFG blend). Defining IR in this way takes into account both the KR and MR of a given VOC species, and, in principle, the incremental reactivity can be broken into kinetic and mechanistic reactivities:

$$IR_i = KR_i \times MR_i \qquad (3\text{-}3)$$

where KR_i and MR_i are, respectively, the kinetic and mechanistic reactivities of the species i.

The IR, as defined by Equation 3-2, is an absolute measure of ozone-forming potential (e.g., the number of grams of ozone per gram of VOC). A somewhat more useful quantity for developing ozone mitigation strategies is the relative incremental reactivity (RIR). RIR is defined as the reactivity of one compound normalized to the reactivity of a base mixture:

$$RIR_i = \frac{IR_i}{\sum_{j=1}^{n} f_j \, IR_j}, \qquad (3\text{-}4)$$

where IR_i is the incremental reactivity of species i, f_j is the fraction of species j in a base mixture containing n different VOCs so that the denominator in the above expression is the total incremental reactivity of a base mixture, such as an RFG blend. The advantages of working with relative incremental reactivities are threefold. First, in a policy-making context, comparisons of reactivities between species or VOC sources are often of most interest. Second, RIR tends to be less sensitive to variations in ambient conditions and thus provides a more robust measure of reactivity. Third, RIR is often easier to develop from three-dimensional models, because there is no apparent absolute scale (e.g., the location and timing of how ozone changes is not uniform) (see McNair et al. 1994).

The two dominant methods that have been used to assess species' reactivities (IR and RIR) are via direct experimental measurement, for example, in an environmental (or smog) chamber, and numerical simulation using computer-based, air-quality models (Carter and Atkinson 1989; Carter 1994; Derwent and Jenkin 1991; Bowman and Seinfeld 1995; McNair et al. 1992; Yang et al. 1995; Bergin et al. 1996). Both methods have serious limitations. Smog chambers do not realistically represent the physics of pollutant transport and the impact of fresh emissions. Moreover, most do not operate over the full range of NO_x

concentrations and VOC to NO_x ratios typically encountered in the polluted atmosphere. Thus, the conditions inside a smog chamber do not reflect those of the ambient air. Given the sensitivity of many VOC reactivities to environmental conditions, smog chamber experiments, by themselves, provide reactivity estimates that are less applicable to atmospheric conditions than those derived from air-quality models. Furthermore, smog chambers have artifacts (e.g., chamber wall and background effects) that can affect the results, particularly if the compound reacts slowly or has radical sinks in its mechanism (Carter and Lurmann 1991). However, chamber experiments are necessary to develop (parameterized) chemical mechanisms for those VOCs for which product and mechanistic data are not yet available from laboratory studies. Data from those chemical mechanisms can then be included in chemical mechanisms for the assessment of their ozone-forming potentials.

Because models can be run for conditions that more accurately reflect actual atmospheric conditions, they can, in principle, provide a more appropriate measure of a species' reactivity than that obtained from a smog chamber. However, virtually all photochemical mechanisms used in current air-quality models are based on data from smog chambers. Thus, the ability of models to accurately simulate air quality depends critically upon reliable extrapolation of smog chamber data to atmospheric conditions and elimination of chamber wall and background effects. This has proven to be a very difficult task (Dodge in press). For these reasons, a level of uncertainty is inherent in any assessment of ozone-forming potential. A variety of approaches has been adopted that attempt to characterize and minimize this uncertainty and thus provide a foundation if reactivity were to be implemented in a policy-making context.

CHEMICAL MECHANISMS AND THEIR DEVELOPMENT

Because of the aforementioned limitations of smog chambers, air-quality models have played a central role in the quantification of VOC reactivity. Of the various components within air-quality models, the chemical mechanism, which attempts to reproduce the VOC-NO_x-air photooxidation process discussed in Chapter 2, is perhaps the most critical component when these models are used to quantify reactivity. This section briefly reviews how these mechanisms are developed and discusses principal mechanisms currently in use.

Any chemical mechanism used in an air-quality model must be designed so that it can, at a minimum, reproduce the major features of the VOC-NO$_x$- air photooxidation process. The principal chemical mechanisms used in current air-quality models, along with representative airshed modeling applications and their key attributes, are listed in Table 3-3. With the exception of the Harwell Master Chemical Mechanism, all the chemical mechanisms in use today include various kinds of parameterizations, approximations, and condensations to simplify the very complex chemical processes that actually occur when VOCs are oxidized in the atmosphere. There are hundreds of different organic compounds in the atmosphere, and from a numerical point of view, it is often impractical to explicitly follow each species. If this were attempted, the chemical mechanisms would be huge (e.g., the Harwell Master Chemical Mechanism (Jenkin et al. 1997) that has over 7,000 reactions) and would be computationally burdensome in three-dimensional models.

TABLE 3-3 Commonly Used Chemical Mechanisms for Air-Quality Modeling and Reactivity Studies

Mechanism	Description	Reference
Statewide Air Pollution Research Center 1990 (SAPRC-90/93/97)	Explicit for a large number of organics, but uses a lumped representation for reactive products. Designed, in part, for reactivity applications.	Carter 1990, 1995, 1997
Carbon Bond IV (CB4)	Lumped by number of carbon bonds in compounds. Specified by EPA for regulatory purposes.	Gery et al. 1989
Lurmann, Carter, Coyner (LCC)	Earlier and more-condensed version of SAPRC-90. Used for the earlier CIT grid-model reactivity-assessment calculations.	Lurmann et al. 1987
Regional Acid Deposition Model, version 2 (RADM-2)	Developed for use in regional acid-deposition modeling. Similar to LCC in detail, except more detailed model for peroxide formation.	Stockwell et al. 1990
Harwell	Used in Europe. Very large number of compounds represented explicitly.	Derwent and Hov 1979
Harwell Master Chemical Mechanism	Detailed, explicit mechanism with over 7,000 reactions.	Jenkin et al. 1997

Moreover, if it were practical, laboratory data are available for only a small subset of the relevant reactions, and for all others their rate constants and the products they form would have to be estimated by extrapolation or by analogy from the simpler, better-studied systems. Thus, preserving the full complexity of the atmospheric VOC chemical system in a model might not necessarily increase the reliability of the model's predictions. Chemical mechanisms in air-quality models, therefore, are typically based on the assumption that the atmospheric oxidation of complex VOCs can be simulated by analogy to simpler ones or by using parameterizations to describe the full suite of elementary reactions. To ensure that these simplifying assumptions are capable of adequately simulating the real world, chemical mechanisms should be, and generally are, tested against experimental data from smog chambers in which the relevant chemical processes are monitored under controlled and well-characterized conditions. These data are then used to tune the various parameterizations contained in the mechanism or to test whether model predictions using the mechanism match experimental results.

Various types of chamber experiments are used to test different aspects of the chemical mechanisms. Irradiations of single VOCs in the presence of NO_x are used to test the mechanism's ability to simulate the oxidation of and ozone production from an individual VOC; NO_x-air irradiations of more complex VOC mixtures test the performance of the model as a whole; and experiments in which the effect of adding single VOCs to irradiations of NO_x and complex mixtures test model predictions of the VOC's incremental reactivity. Evaluation of chemical mechanisms with smog chamber data is complicated by uncertainties in chamber effects, and separate characterization experiments are needed to evaluate those effects.

Chamber data are currently available to test the mechanisms for only a subset of the many types of VOCs emitted into the atmosphere. For the other species, reactions are either derived by analogy with mechanisms for compounds that have been studied, or they are represented in the model as if they reacted in the same way as some other chemically similar species. Mechanisms are further simplified or extrapolated using an approach referred to as "lumping." In this approach, a single hypothetical (or pseudo) species is used in the model to represent a larger number of compounds assumed to react in the same way, or a group of model species is used to represent aspects of the reactions of various chemical compounds. The lumping approaches, and the approximations and inaccuracies they introduce, vary depending on the mechanisms (see Table 3-3).

REACTIVITY ASSESSMENTS USING SMOG CHAMBERS

As mentioned above, one way to assess a VOC's reactivity is to measure its effect on ozone when irradiated in the presence of NO_x and other VOCs in smog-chamber experiments. Although these results have limited applicability for the reasons discussed above, they can be quite valuable for evaluating and verifying reactivities calculated using air-quality models. Studies based on smog chambers include those of Carter at the University of California at Riverside, Kelly at the General Motors Research Laboratories, and Jeffries at the University of North Carolina. The results of the experiments have been encouraging.

Carter and Atkinson (1987) conducted a series of experiments in which the impact of adding a VOC to a base mixture of organics and NO_x was compared with a similar experiment without the extra compound being added. This series was done at various NO_x levels. Results of those and more recent series of experiments have been compared with the predictions of both the SAPRC-90 mechanism and SAPRC-93 mechanism. The reactivities calculated using the SAPRC-90 mechanism agreed reasonably well with the experimental results for most VOCs, except for the internal alkenes (e.g., 2-butene, 2-pentene). Reactivities calculated using the SAPRC-93 mechanism performed significantly better. In particular, the mechanisms performed quite well in simulating the effects of varying the NO_x levels and the nature of the reactive VOC surrogate. However, neither mechanism performed particularly well in simulating reactivity differences among xylene and trimethylbenzene isomers.

In the experiments of Kelly et al. (1994, 1996), incremental reactivities of several representative VOCs were measured as a function of the amount of VOC added under conditions that tended to maximize the reactivity. Although the VOC mix used in the experiments only approximated the VOC mix simulated in a replicate modeling study, the experimental reactivity results correlated well with the modeled reactivity results.

Jeffries et al. (1997, 1998) used a large outdoor smog chamber to study ozone formation from various complex mixtures designed to closely duplicate components in vehicle exhausts, and Kleindienst et al.

(1996) performed similar experiments using an indoor smog chamber to examine the reactivity of the exhaust from vehicles using alternative fuels. The purpose of the Jeffries et al. studies was to evaluate chemical mechanisms, and to compare, directly, ozone formation from various chemically realistic mixtures.

Although smog-chamber studies are essential for chemical-mechanism evaluation, incremental reactivities in smog chambers are not the same as incremental reactivities in the atmosphere (as discussed above). It is not practical to duplicate all the environmental conditions that affect a VOC's incremental reactivity in smog-chamber experiments, and, even if it were practical to do so, it would not be practical to use such information to investigate comprehensively how reactivities vary over the wide variety of conditions that occur in the atmosphere. For this, air-quality model calculations are required.

AIR-QUALITY MODELS

Air-quality models are computerized representations of the atmospheric processes responsible for air pollution, which includes ozone formation (NRC 1991), These models integrate current understanding of the atmosphere's chemistry and meteorology with estimates of source emissions to predict how the composition of trace atmospheric species, such as ozone, respond to changes in emissions. Table 3-4 lists and describes some of the air-quality models that have been used to assess VOC reactivity and the ozone-forming potential of motor-vehicle emissions.

The models vary greatly in complexity, and thus also vary in the amount of input data and computational resources they require. In general, the major processes that affect the evolution of pollutants are parameterized within the models, including emissions releases, gas-phase chemical reactions (using chemical mechanisms as described above), transport, mixing, deposition, and scavenging. The equation upon which air-quality models are founded is a statement of chemical species conservation (Seinfeld 1986):

$$\frac{\partial c_i}{\partial t} + \nabla \bullet \overline{U} c_i = \nabla K \nabla (c_i / r) + R_i (c_1, \ c_2, \ \ldots \ c_n, \ T, \ t) +$$

$$S_i (\overline{x}, \ t) \ i = 1, \ 2, \ 3, \ \ldots, \ n \qquad (3\text{-}5)$$

TABLE 3-4 Examples of Air-Quality Models[a]

Model	Reference	Description
Empirical Kinetic Modeling Approach (EKMA)	Dodge 1977; Gipson 1984	Lagrangian, single well-mixed cell. Allows for time-varying emissions and inversion height raise.
Urban Airshed Model (UAM)	Reynolds et al. 1973, 1979	Three-dimensional, urban-scale photochemical model. Specified by the EPA for regulatory applications.
Carnegie/California Institute of Technology (CIT)	McRae et al. 1982; Harley et al. 1993	Three-dimensional, urban-scale photochemical model.
CALGRID	Yamartino et al. 1989, 1992	Three-dimensional, urban-scale photochemical model.
Regional Oxidant Model (ROM)	Lamb 1983	Three-dimensional, regional-scale photochemical model.
Urban-to-Regional Multi-scale (URM) Model	Odman et al. 1994	Three-dimensional, multiscale photochemical model.
EPA Models-3	Dennis et al. 1996	Three-dimensional, multiscale photochemical model.

[a]For more information on types of air-quality models and model verification, see Russell and Dennis 1998, and the references therein.

where, on the left,

$\dfrac{\partial c_i}{\partial t}$ represents the local time rate of change in c_i, the concentration of species i, and

$\overline{U}c_i$ represents the rate of transport of species i by organized wind fields (i.e., advection); on the right,

$K\nabla c_i$ represents the rate of transport due to turbulent mixing,

R_i is the net rate of change in c_1 through c_n due to chemical reactions for time t and temperature T, and

S_i represents emissions (sources) of compound i over a specified time and at a specified location.

The differences in air-quality models arise primarily from the varying degrees of complexity allowed in the treatment of the nonchemical processes and in the numerical techniques used to solve Equation 3-5. To date, model simulations of ozone formation and VOC reactivity have been performed using two types of air-quality models: (1) box or trajectory models and (2) three-dimensional Eulerian models.

The trajectory or box model represents the polluted atmosphere by a discrete air parcel. (This model is the kind used to illustrate aspects of ozone chemistry in Chapter 2.) Many trajectory models use a single cell to represent a column of boundary-layer air; others use discreet cells to subdivide the vertical column (e.g., the two-cell model used by Derwent and Jenkin 1991). The model's air parcel either is fixed in space (i.e., as a box over a city) or allowed to move over the air basin, following a trajectory calculated from the wind fields (i.e., a Lagrangian simulation). In either case, emissions, deposition, and meteorological changes can be included. However, box and trajectory models, by their very nature, greatly simplify transport and diffusion, provide very limited information on spatial variability, and thus cannot represent any particular pollution episode with great detail. However, they can represent chemical transformations in as great detail as is known. Further, they are readily applied and are not computationally intensive. For these reasons, box models have been used extensively to define reactivities. For example, the reactivity scale specified by the California Air Resources Board (CARB) in the California LEV/CF Regulations (CARB 1990) was developed using a single-cell model (see discussion below). To test how well these models predict reactivity in a specific airshed, and to examine the spatial and temporal aspects of VOC reactivity, a more physically detailed Eulerian model must also be applied.

Three-dimensional Eulerian models, also called grid or airshed models, divide a represented air mass into multiple vertical and horizontal cells. Grid models provide the most comprehensive representation of any airshed and provide the only means to predict observed pollution levels in real-world pollution episodes, particularly with respect to spatial and temporal variation. However, these models require large quantities of detailed input data and have relatively high computational demands. In addition to uncertainties in chemical mechanisms (a feature common to both box and grid models), grid models are also often limited by uncertainties in input data (e.g., emissions and wind fields). For these reasons, grid models are best applied to airsheds in which extensive, carefully examined input data are available. Results can then be compared with ambient pollutant observations to evaluate the accuracy of model predictions. Although models are frequently only evaluated against observed ozone data, comparisons with observations of VOC and NO_x concentrations are needed to assess a model's ability to accurately simulate the relationships between ozone and its precursor emissions.

Thus, box and grid models provide varying advantages and disadvantages. Because they are not computationally intensive, box models can be used to represent a wide variety of chemical conditions and to perform extensive, formal sensitivity analyses. Grid models, on the other hand, although not well suited to multiple scenario testing and comprehensive sensitivity analysis, provide an opportunity to comprehensively assess specific pollution scenarios with great spatial and temporal detail. Choosing which model is best suited for a specific application is often based on balancing the need for physical detail with computational limitations. For these reasons, the study of reactivity should, in principle, rely on both box- and grid-model predictions. In this case, results from both types of models can be compared to help assess the reliability of the predictions. Air-quality modeling studies conducted specifically for investigating VOC species reactivity are given in Table 3-5.

Box- and Trajectory-Model Reactivity Assessments

Carter and Atkinson (1989) used a box model and a detailed chemical mechanism to quantify the reactivities of a variety of VOCs. They found, not surprisingly, that the reactivity in terms of grams of ozone per grams of VOC varied significantly between compounds and also as a function of the VOC to NO_x ratio. In follow-on work, Carter (1993, 1994) developed 18 separate reactivity scales for quantifying VOC reactivity under different conditions, in this case using the SAPRC-90 chemical mechanism in a single-cell trajectory model. Those reactivity scales were derived using nine different approaches for dealing with the dependence of reactivity on environmental conditions and on two methods for quantifying ozone impacts. Of the 18 reactivity scales, 3 have received the most attention: the maximum incremental reactivity (MIR) scale, the maximum ozone incremental reactivity (MOIR) scale, and the equal benefit incremental reactivity (EBIR) scale (see Table 3-6).

The MIR scale is the incremental reactivity (IR) of a VOC computed for conditions in which the compound has its maximum absolute IR value. This generally occurs at a low VOC-to-NO_x ratio in which the

TABLE 3-5 Examples of Compound-Reactivity Modeling Studies

Reference	Model Type	Mechanism	Application
Carter and Atkinson 1989	Trajectory	SAPRC	One-day simulation of reactivities under varying VOC-NO$_x$ conditions.
Derwent and Jenkin 1991	Trajectory	Derwent and Hov (1979)	Two-layer 5-day trajectory simulations of reactivity. Photochemical ozone creation potential (POCP) scales.
McNair et al. 1992	Three-dimensional (CIT)	LCC	Calculation of three reactivity scales for 11 lumped compounds. Simulations were performed for a 3-day period in the Los Angeles area (the SCAQS episode).
Carter 1994	Trajectory	SAPRC-90	Development of 18 reactivity scales (including the maximum incremental reactivity (MIR) and the maximum ozone incremental reactivity (MOIR)) for 117 compounds. Results are the average of 39 trajectory simulations for 10-hr periods.
Yang et al. 1994	Trajectory and three-dimensional (CIT)	SAPRC-90	Review of rate constant uncertainties
Yang et al. 1995	Trajectory	SAPRC-90	Rate constant uncertainty calculations for the reactivities of 26 compounds under MIR- and MOIR-type conditions. One averaged trajectory was used rather than the 39 used in the Carter MIR and MOIR calculations.
Bergin et al. 1995	Three-dimensional (CIT)	SAPRC-90	Calculation of three reactivity scales for 27 compounds. Simulations were performed for the SCAQS episode.
Jiang et al. 1996	Trajectory	SAPRC-90	Calculation of the contributions of 18 compounds to ozone concentrations in the Lower Fraser Valley.
Derwent et al. 1996	Harwell trajectory model	Harwell mechanism	Updated calculation of VOC POCPs.

Table 3-5 *(Continued)*

Reference	Model Type	Mechanism	Application
Bergin et al. 1998a	Three-dimensional (CIT)	SAPRC-90	Rate constant uncertainty calculations for the scales and compounds in the Bergin et al. (1995) study above.
Derwent et al. 1998	Harwell trajectory model	Master Chemical Mechanism	Calculation of VOC POCPs using a large, detailed mechanism.
Khan et al. 1999	Trajectory and three-dimensional	SAPRC-90	Calculation of eight compound reactivities in three domains using both grid and box modeling.

chemistry is VOC-limited (see Figure 3-1). Mathematically, it is approximately expressed as

$$MIR_i = \text{Average}\left\{ \text{Max}\left[\frac{\partial[O_3]}{\partial[VOC_i]} \right] \right\} \qquad (3\text{-}6)$$

where MIR_i is the MIR of species i, and $[VOC_i]$ is the input amount of species i. In practice, Carter fixed the VOC concentrations and adjusted the NO_x to maximize the reactivity for the specific model run (or trajectory). The MOIR scale is the incremental reactivity computed for conditions that maximize the ozone concentration (see Figure 3-1), and thus tends to represent conditions in which the VOC to NO_x ratio is moderate and the chemistry is approaching, or in, the transitional region between VOC limitation and NO_x limitation (see Chapter 2). Mathematically, it is

$$MOIR_i = \text{Average}\left\{ \left[\frac{\partial[O_3]}{\partial[VOC_i]} \right]_{\text{for } [O_3] \text{ maximized}} \right\} \qquad (3\text{-}7)$$

In this case, the NO_x levels in the trajectories are typically set to give the maximum ozone levels, and then the sensitivity of the ozone to the

TABLE 3-6 Summary of Major Characteristics of the Primary Carter Reactivity Scales

Scale	Type of Scenarios Used	Derivation of Scale from Individual Scenario Reactivities	Ozone Quantification	Reflects Effect of VOC on
Maximum incremental reactivity (MIR)	Low VOC-to-NO$_x$ ratio conditions in which ozone is most sensitive to VOC changes	Averages of incremental reactivities in the MIR scenarios	Maximum ozone	Ozone formation rates
Maximum ozone incremental reactivity (MOIR)[a]	Moderate VOC-to-NO$_x$ ratio conditions in which highest ozone yields are formed	Averages of incremental reactivities in the MOIR scenarios	Maximum ozone	Ultimate ozone yield
Equal benefit incremental reactivity (EBIR)	Higher VOC-to-NO$_x$ ratio conditions in which VOC and NO$_x$ control are equally effective in reducing ozone	Averages of incremental reactivities in the EBIR scenarios	Maximum ozone	Ultimate ozone yield

[a]The MOIR scale is also referred to as the maximum ozone reactivity (MOR) scale.

individual VOCs is assessed. EBIR is the incremental reactivity for the conditions in which the sensitivity of ozone to VOC is equal to that of NO$_x$. Thus, the EBIR scale is calculated for conditions that lie midway between VOC limitation and NO$_x$ limitation (i.e., the transitional regime).

CARB (1990) proposed using the MIR scale for regulatory applications, because the MIR scale reflects reactivities under environmental conditions that are most sensitive to the effects of VOC controls. Although the MIR scale might not be accurate for lower NO$_x$ conditions, State of California officials reasoned that, because of the lower sensitivity of ozone to VOC under these conditions, the impact of these inaccuracies would not be as critical (i.e., the scale would be most accurate for VOC-limited conditions, the conditions for which VOC controls would be most effective). The MIR scale was also found to correlate well to scales based on integrated ozone yields, even in lower NO$_x$ scenarios. Perhaps for

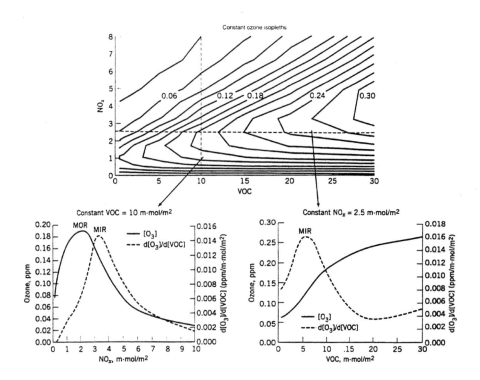

FIGURE 3-1 Dependencies of peak ozone concentrations and the peak ozone sensitivities ($\partial[O_3]/\partial[VOC]$) with respect to initial VOC and NO_x concentrations. The top graph illustrates peak ozone concentrations (as isopleths) as a function of both VOC and NO_x. The bottom left hand graph shows how peak ozone levels vary when NO_x is increased at a constant VOC input, and the right hand graph shows how ozone changes as VOC is varied at constant NO_x input. Also shown is how the sensitivity ($\partial[O_3]/\partial[VOC]$) varies. The peak in the ozone sensitivity ($\partial[O_3]/\partial[VOC]$) plot corresponds to MIR conditions (in essence, maximum sensitivity), and the peak in the ozone concentration plot corresponds to MOIR (i.e., maximum ozone) conditions. The maximum ozone concentrations were calculated using a 1-day box-model simulation using the averaged conditions scenario of Carter (1994) and the SAPRC-93 mechanism. Source: Bergin et al. 1998a. Reprinted with permission from *Encyclopedia of Environmental Analysis and Remediation*, copyright 1998, John Wiley & Sons, New York.

these reasons, MIR has been the reactivity scale used most extensively for policy-making in the United States. For example, in California, the MIR scale is used as a basis for deriving reactivity adjustment factors (RAFs)[4] in California's LEV/CF regulations (CARB 1991). The MIR scale was also used to compare reactivities of vehicle emissions during various driving cycles as well as with the use of various reformulated gasolines in the Auto/Oil Study sponsored by the petroleum and automobile manufacturing industries (AQIRP 1993). Thus, the analyses presented later in this report are also largely based on the MIR scale.

Nevertheless, it should be noted that the MOIR and EBIR scales have advantages. For example, MOIR is representative of conditions for the worst case scenario in which ozone concentrations would be highest. Both MOIR and EBIR are more representative of lower NO_x conditions that are typically found in the eastern United States. Moreover, the MIR scale tends to predict lower reactivities for slowly reacting compounds than might be appropriate, because the higher NO_x concentrations used for MIR scenarios tend to suppress radical levels and thus also suppress the kinetic reactivity of slower-reacting compounds.

Other trajectory-model investigations of VOC reactivity have included Andersson-Skold et al. (1992), Derwent and Jenkin (1991), and Derwent et al. (1996, 1998). Those researchers derived a comparable set of VOC reactivities, termed photochemical ozone creation potentials (POCPs). POCP is defined as the reactivity normalized to ethene calculated using a two-layer trajectory model covering an idealized 5-day trajectory across Europe. The second layer contains reacted material from previous days. The POCPs are calculated from the change in mid-afternoon ozone concentration due to each species in the trajectory that results from removing the test VOC from the emissions, divided by the integrated emissions of the test VOC up to the time of the ozone observation.

A comparison of MIR, MOIR, and POCP for selected VOCs is shown in Figure 3-2. The MIR and MOIR scales usually give similar relative reactivities for most compounds, and are consistent in their predictions of which compounds are highly reactive and which are not. However, for reasons indicated above, the MOIR scale gives lower relative reactivities for aromatics, and also predicts lower relative reactivities for radical

---

[4]RAF (reactivity adjustment factor) is the ratio between the exhaust reactivities of two fuels (see discussion later in this chapter).

54

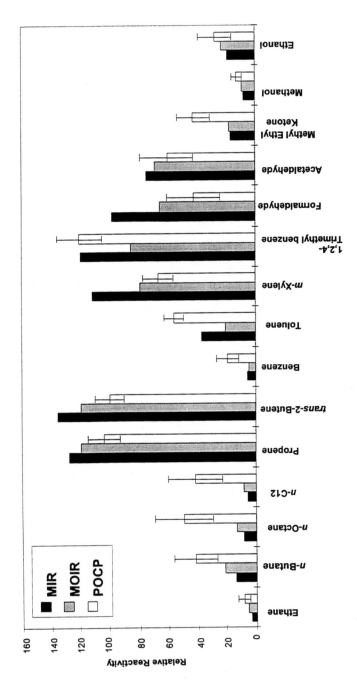

FIGURE 3-2 Comparison of MIR, MOIR, and POCP for selected VOCs. (Incremented reactivities (MIR and MOIR) are normalized relative to ethane = 100. POCP reactivities are averages for various trajectories. Error bias represent standard deviation of averages. Source: Bergin et al. 1998a. Reprinted with permission from *Encyclopedia of Environmental Analysis and Remediation*, copyright 1998, John Wiley & Sons, New York.

initiators, such as formaldehyde, which have larger effects on rates of ozone formation than on total ozone formation over longer periods. Effects of differences and uncertainties in chemical mechanisms on calculated incremental-reactivity scales are discussed in more detail later in this chapter.

Eulerian-Model Reactivity Assessments

A serious concern about the regulatory application of scales, such as MIR and MOIR, is that they are based on a box-model or trajectory-model simulation of a single-day air-pollution episode. For example, although MIRs are often developed from 10-hr simulations, some organic compounds can remain in an urban airshed for 2 to 3 days if stagnation is sufficiently severe or there is significant recirculation. Thus, MIRs might underestimate the relative reactivity of the slower-reacting compounds. Moreover, trajectory models lack the physical detail, the spatial and temporal detail of emissions and resulting pollutants, and the multiday pollution effects that can be represented in Eulerian models. For that reason, reactivities derived using box and trajectory models should ideally be evaluated using more-detailed Eulerian models. On the other hand, such an evaluation is not without its own inherent challenges. One of the most crucial is establishing a protocol for comparing model results; that is, what aspect of the spatially and temporally detailed Eulerian-model predictions are most appropriate to compare with a relatively simple MIR or MOIR predicted by a trajectory model? Perhaps somewhat arbitrarily, investigators have typically used either the Eulerian-model predicted values for the peak ozone concentration in the airshed or the population-weighted exposures to ozone.

Thus far, the most comprehensive comparison of reactivities calculated using trajectory models with those derived from an Eulerian model have been carried out using the Carnegie/California Institute of Technology (CIT) model (Harley et al. 1992) applied to a 3-day air-pollution episode in the Los Angeles air basin (McNair et al. 1992; Bergin et al. 1995, 1998b; Khan et al. 1999). McNair et al. (1992) used the CIT model with a highly lumped chemical mechanism (the Lurmann et al. (1987) mechanism (LCC)) to quantify the reactivity of 11 individual and lumped VOCs. This study allowed comparison with single-cell-model reactivity studies by others; it also allowed comparison of the different metrics used to derive reactivities. The results showed that MIRs derived from trajectory models did not perform well in predicting peak ozone sensitivities to specific VOC species, but performed reasonably well in predicting the effects of VOC species on the integrated exposure to ozone over the air-quality standard. The MOIR scale did not compare as well as the MIR scale with results derived from airshed model for either the peak ozone concentration or ozone exposure concentrations greater than the air-quality standard.

Subsequent to the study of McNair et al. (1992), the SAPRC-90 mechanism was implemented in the CIT model by Bergin et al. (1995, 1998a) for more direct comparison with the MIR and MOIR scales. Reactivities were normalized to a mixture of VOCs representative of exhaust emissions, as in the reactivity studies of Carter (1994) and Yang et al. (1996). Again, the results for the exposure metrics compared well with the MIR scale (e.g., regression gave a slope of 0.98 and $r^2 = 0.97$). To a lesser extent, the MOIR scale compared reasonably well with the peak ozone metric from three-dimensional modeling (slope = 0.95, $r^2 = 0.74$), which occurs in a region that is less NO_x-rich. These results suggest that the MIR scale is most appropriate in areas rich in NO_x, though is less well suited to areas that are more NO_x poor. This is examined further in the discussion on variabilities.

UNCERTAINTIES IN SPECIES' REACTIVITIES DUE TO CHEMICAL-MECHANISM UNCERTAINTY

A concern often raised with regard to the use of reactivities in policy-making is their dependence on model-derived quantities that might be significantly distorted by uncertainties in knowledge of atmospheric chemistry and its representation through chemical mechanisms. Measurement errors in laboratory kinetic and product studies contribute to uncertainty in the chemical mechanisms used to calculate incremental reactivities. Moreover, as discussed above, the products of the initial OH radical, NO_3 radical and/or ozone reactions, and their subsequent products, of many of the organic compounds emitted into urban atmospheres are not well characterized. Their representation in chemical mechanisms is based on analogy to compounds of similar structure, creating added uncertainty. At issue is whether the uncertainties in the chemistry, not only of the target species but others present in the atmosphere as well, significantly limit the reliability of model-derived reactivities for organic compounds. The impact of uncertainties in chemical mechanisms on the reliability of reactivities derived from models should be discussed at two levels. First, how uncertainties affect the reactivity of individual VOCs is addressed in this section. Second, how they affect the reactivity of a source of emissions whose composition is made up of a large number of VOCs is addressed later in this chapter with particular emphasis on light-duty vehicular (LDV) emissions.

One way to assess the effects of chemical-mechanism uncertainty is to compare reactivity predictions using different state-of-the-art mechanisms that incorporate differing assumptions concerning unknown areas of the chemistry and differing lumping approaches. As discussed above, the SAPRC-90 mechanism was used for calculation of the MIR, MOIR, and other reactivity scales because of the large number of VOCs it can explicitly represent. The RADM-2 and LCC mechanisms employ assumptions similar to SAPRC-90 concerning uncertain portions of the aromatics and other mechanisms, and would be expected to give similar reactivities for the species that the condensed mechanisms are designed to represent. However, this might not be the case for the CB4 mechanism, which employs different assumptions concerning some of the uncertainties in the aromatics mechanisms, and uses different methods for treating alkane and alkene reactions (Gery et al. 1988). In addition, since the CB4 mechanism and SAPRC-90 mechanism were developed, there have been significant changes in the understanding of alkene and ozone reactions, new data on aromatics mechanisms, new laboratory data concerning a number of potentially important reactions, and a large database of new smog-chamber experiments designed explicitly to test VOC-reactivity scales (Carter et al. 1993; Jeffries and Sexton 1995; Carter et al. 1995a,b,c).

Figure 3-3 shows a comparison of MOIRs and MIRs for vehicular exhaust emissions (relative to standard exhaust) calculated with the SAPRC-90, CB4, and the updated SAPRC-93 mechanisms. Other than the mechanism, the scenarios and the calculation methodology were the same (Carter 1994). Differences of about 20% or more are not uncommon. However, for ethanol and MTBE, the agreement among the mechanisms is remarkable. The most conspicuous difference is for toluene.

More systematic studies of the effects of mechanism uncertainties were carried out by researchers (Derwent and Hov 1988; Russell et al. 1995; Yang et al. 1995, 1996; Bergin et al. 1996, 1998a; Yang and

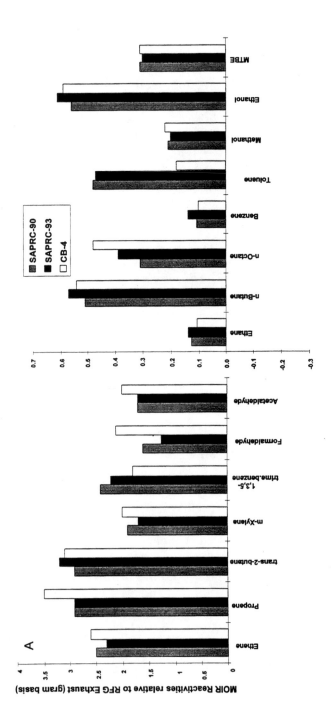

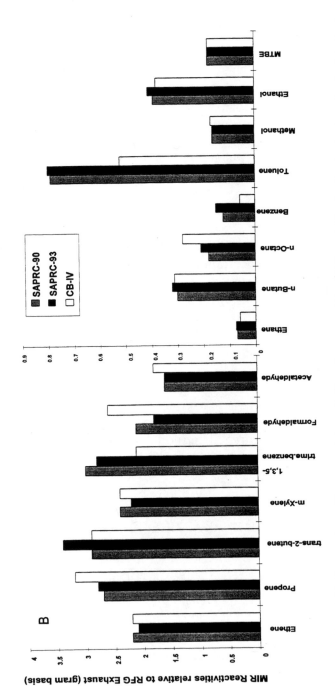

FIGURE 3-3 (A and B) Comparison of incremental reactivities of represenative VOCs, relative to standard exhaust calculated using the SAPRC-90, SAPRC-93, and Carbon Bond IV mechanisms. Source: Bergin et al. 1998a. Reprinted with permission from *Encyclopedia of Environmental Analysis and Remediation*, copyright 1998, John Wiley & Sons, New York.

Milford 1996) using airshed models and box models to explore to what degree uncertainties in chemical-rate parameters affect the calculated compound reactivities. Yang et al. (1995, 1996) used Monte Carlo analysis with Latin Hypercube Sampling to calculate reactivity uncertainties derived from a trajectory model. Bergin et al. (1998a) extended this analysis to a three-dimensional model by focusing on only those uncertainties in the chemical mechanism identified by Yang et al. to be most critical. Generally, these studies suggest that the uncertainty[5] in the mean MIR value calculated for most individual VOCs generally is in the range of 20% to 60%. The estimated uncertainty in the predicted peak ozone concentration for the average MIR simulation conditions was about 30%, relative to a mean prediction of ~0.15 ppm. For predicted ozone and MIRs, the most influential uncertainties are those in rate parameters that control the availability of NO_x and radicals (Yang et al. 1995). For MIRs, uncertainties in the rate parameters of primary oxidation reactions, or reactions of relatively stable intermediates, are also influential. However, because uncertainties in the rate constants and parameterizations used in the chemical mechanisms apply to the calculations for all VOC reactivities, the effects of these uncertainties on the reactivities of individual VOCs are strongly correlated between VOCs. For example, an increase in the photolysis rate for NO_2 increases the reactivity of most species by about the same proportion. Thus, the relative reactivities tend to have significantly smaller uncertainties than those of the absolute reactivities (Yang et al. 1995, 1996). Generally, through the use of three-dimensional modeling, the uncertainties in the relative reactivities of individual VOCs have been found to range from about 15% to 40% (Bergin et al. 1998a).

VARIABILITY OF OZONE-FORMING POTENTIAL WITH ENVIRONMENTAL CONDITIONS

Another concern about the use of reactivities within a regulatory context

---

[5]In this and subsequent sections, uncertainty denotes two times the standard error of the mean. Such confidence intervals will contain the actual value 95% of the time. A more-detailed discussion of uncertainty and its implications for policy-makers is presented in Chapter 7.

relates to the fact that the ozone-forming potential of any given VOC can be heavily dependent upon local ambient conditions. In the extreme, a compound can go from being an efficient generator of ozone under one set of conditions to having a negative impact on ozone production under another set of conditions. This is due, in part, to the formation of an organic nitrate that ties up both a photochemically active oxidized nitrogen molecule and a reactive organic radical. While some compounds (e.g., toluene) do appear to have this property, a variety of studies indicate that such compounds represent exceptions rather than the rule, and that, as in the case of mechanistic uncertainty, the impact of environmental variability can be minimized by using relative reactivities instead of absolute reactivities. A few studies that have addressed those complications are discussed below.

In order to assess the magnitude of reactivity variability from one city to another, Russell et al. (1995) derived absolute and relative reactivities along 39 trajectories using the box model of Carter (1994). Mean absolute reactivities and mean relative reactivities, along with their respective standard deviations of the mean, were then calculated. The magnitude of those standard deviations thus provides an indication of how different environmental conditions affect reactivities. Inspection of Table 3-7, in which some of the standard deviations calculated by Russell et al. are listed, indicates that environmental variability does in fact introduce significant variability into reactivities for many of the ubiquitous VOCs. However, such variability can be reduced by almost of factor of 2, from about 25-60% to 15-40% through the use of relative reactivities instead of absolute reactivities.

TABLE 3-7   Uncertainty in the Mean Absolute and Relative MIRs from 39 Separate Trajectory Simulations Representing Different Environmental Conditions

| Compound | 95% Confidence Interval (% of Mean Value) | |
	Absolute Reactivity	Relative Reactivity
Formaldehyde	28	16
Methanol	39	23
Ethane	56	38
Toluene	38	21
Pentene	39	21

Source: Derived from Russell et al. 1995.

An alternate approach, adopted by Bergin et al. (1995, 1998a), assessed the impact of environmental variability by comparing reactivities calculated using a three-dimensional, grid-based model for the Los Angeles area with those derived from box-model simulations for 30 cities. Because of their large spatial domain, three-dimensional models cover domains with a wide range of environmental conditions and the reactivities derived from these models represent composite averages over this domain. Reactivities from box models, on the other hand, focus on a single set of environment conditions corresponding to a specific air mass following a specific trajectory. In Figure 3-4 the relative reactivities for a variety of VOCs calculated on the basis of peak ozone concentrations, population-weighted ozone exposures, and spatially weighted ozone exposures from a three-dimensional-model simulation are compared along with box-model-derived results. Here again, while significant variability is seen (and in the cases of toluene and ethylbenzene a change in the sign of the relative reactivity), the general trend in the reactivities from one species to another tends to be reasonably consistent.

Another relevant study is that of Khan et al. (1999), who conducted a reactivity study on eight VOC solvents having a wide range of reactivities in three different airsheds: Los Angeles, the Swiss Plateau, and Mexico City. Although the relative reactivities for the eight compounds were found to be similar between the Los Angeles and Switzerland domains, the very high VOC loadings found in Mexico City led to more substantial differences, one being that the aromatic species could have negative reactivities. The reactivity of the aromatics being greatly reduced in regions with lower NO_x and higher VOCs was discussed earlier.

UNCERTAINTIES IN RELATIVE REACTIVITIES OF MOTOR-VEHICLE EMISSIONS

The previous discussion pertains to the reactivities of individual compounds and their attendant uncertainties. However, the charge to this committee (see Chapter 1) is to look at the use of relative reactivities as applied to motor-vehicle emissions, which are composed of hundreds of compounds. This complexity introduces some extra issues, in particular emission-composition uncertainty. A variety of modeling studies, listed in Table 3-8, have examined the reactivity of source emissions. In large

63

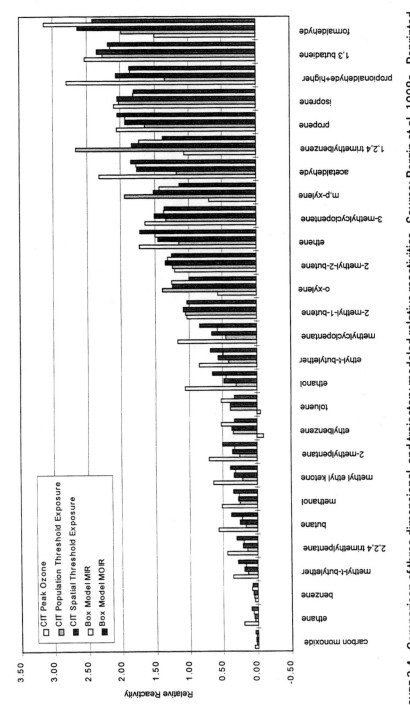

FIGURE 3-4 Comparison of three-dimensional- and trajectory-modeled relative reactivities. Source: Bergin et al. 1998a. Reprinted with permission from *Encyclopedia of Environmental Analysis and Remediation*, copyright 1998, John Wiley & Sons, New York.

TABLE 3-8 Summary of Source Emissions Reactivity Modeling Studies

Reference	Model Type	Mechanism	Application
Trijonis and Arledge (1976)	Calculated (not modeled)	EPA smog chamber data	Estimated major source reactivities for metropolitan Los Angeles.
Chang et al. (1989)	Trajectory	LCC	Methanol-fueled vehicle impacts with respect to conventionally fueled vehicles.
Russell et al. (1990)	Three-dimensional (CIT)	LCC	Potential methanol-fueled vehicle impacts for the SCAQS episode (compared with equal mass emissions from conventional vehicles).
McNair et al. (1994)	Three-dimensional (CIT)	LCC	Calculations of RAFs for four fuels. Simulations were performed for the SCAQS episode.
Yang et al. (1996)	Trajectory	SAPRC-90	Rate constant and exhaust composition uncertainty calculations for the RAFs from reformulated gasolines and methanol.
Bergin et al. (1996)	Trajectory and three-dimensional (CIT)	SAPRC-90	Report on box model study described above and a three-dimensional study of the effects of rate constant and product yield uncertainties on predicted ozone impacts of five alternative fuel RAFs.
Russell et al. (1995)	Trajectory and three-dimensional (CIT)	SAPRC-90	Evaluation of combined results of most previous studies. An economic analysis was also performed.
Dunker et al. (1996)	Three-dimensional (UAM)	CB4	Extensive evaluation of how reformulated and alternative fuels would affect ozone formation in Los Angeles, New York, and Dallas. Tied directly to program to assess how fuel blends affect both emissions composition and emissions rates.
Guthrie et al. (1996)	Three-dimensional (UAM)	CB4	Modeling of potential impacts of the use of three alternative fuels (CNG, M85, and RFG) in two urban areas.

part, because of the alternative fuel regulations promulgated in California (see Chapter 5), this issue has been explored in most detail for motor-vehicle exhaust emissions.

When CARB implemented regulations for the LEV/CF[6] program, it introduced the concept of reactivity adjustment factors (RAFs) to provide a mechanism for manufacturers who build vehicles powered by alternative fuels (including reformulated gasoline) to take advantage of the lower ozone-forming potential of the emissions from these vehicles. An RAF is defined as the ratio of the specific exhaust reactivities of two fuels (per gram of emission of an alternatively-fueled vehicle to that of a conventionally fueled vehicle). The specific reactivity of fuel i (SR_i) is

$$SR_i = \sum_{i=1}^{N} F_{Ai} R_i , \qquad (3\text{-}8)$$

where F_{Ai} is the fraction of species i in fuel A and R_i is the MIR of species i. The RAF for fuel A is defined as the ratio of the exhaust reactivities:

$$RAF = \frac{\displaystyle\sum_{i=1}^{N} F_{Ai} R_i}{\displaystyle\sum_{i=1}^{N} F_{Bi} R_i} , \qquad (3\text{-}9)$$

where F_{Bi} is the fraction of species i in the base (reference) fuel. If the alternative fuel's RAF is less than 1, then a proportionally greater amount of VOCs can be emitted, such that the RAF times the mass of emissions meets some total emissions standard. In practice, the appropriateness of RAF—calculated using MIR values—was tested using a grid model and adjustments were made as necessary.

The sources and magnitude of the uncertainties in RAFs have been investigated by a variety of investigators, including Yang et al. (1996), McBride et al. (1997), and AQIRP (discussed later). The studies of Yang

---

[6]Aspects of this program are discussed in Chapters 4 and 5.

et al. (1995) and McBride et al. (1997) revealed that although the 2-σ uncertainty in the relative reactivity of individual species due to uncertainties in chemical mechanisms generally range from about 20% to 40%, that range grossly overstates the uncertainty in the composite relative reactivity of a specific emissions source. An example would be reactivities from a fleet of motor vehicles using one type of fuel versus another. In this case, much of the chemical uncertainty tends to cancel out (provided one is using relative reactivities instead of absolute reactivities), leaving an uncertainty of only a few percent. A much larger uncertainty arises from the variability and difficulty in characterizing how different vehicles respond to fuel composition changes. This is largely due to the limited amount of test data and the limited knowledge of how well a vehicle fleet is characterized by the data. This leads to substantial uncertainties in the composition of the emissions, which feed directly into the calculation of the source reactivity. The result is an uncertainty (95% confidence level) in relative reactivities for source categories such as motor-vehicle emissions of about 15-30% (Yang and Milford 1996; Bergin et al. 1998a).

REACTIVITY FOR 1-HR PEAK AND 8-HR AVERAGED OZONE CONCENTRATIONS

Another specific question under consideration is whether reactivity scales developed for a peak 1-hr ozone concentration (i.e., in accordance with the current form of the National Ambient Air Quality Standards (NAAQS)) would be significantly different from a similar scale developed for a peak 8-hr ozone concentration (i.e., the new form of NAAQS). At present there is little information to assess this issue. Of relevance is a study of Khan et al. (1999) in which the authors compared the reactivities of several compounds based on their impact on the peak 1-hr and the average 8-hr ozone concentrations. The comparison is shown in Figure 3-5. Major differences were only found in the halogenated aromatics that had very small reactivities to begin with. The relative reactivities of the other species did not change appreciably. This result, albeit limited, appears to suggest that reactivity scales derived for peaks of 1-hr averaged ozone concentration should largely apply to peaks of 8-hr averaged ozone concentrations in urban areas.

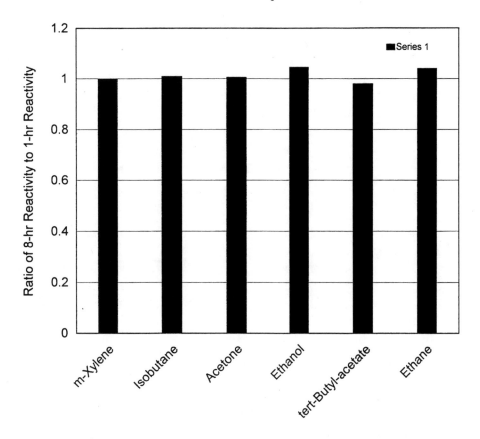

FIGURE 3-5 Ratio of 8-hr average peak ozone relative reactivity to 1-hr average peak ozone relative reactivity for six solvents. Results are for a 3-day simulation in Los Angeles. Source: Adapted from Khan et al. 1999.

On the other hand, a number of caveats should be borne in mind before this result is used to justify the application of trajectory-model-derived reactivity scales based on VOC impact on peak 8-hr averaged ozone concentration. In the first place, recall that Eulierian-model-derived reactivities based on the model's predicted peak ozone concentration did not compare well with the trajectory-model-derived MIRs. Second, reactivities derived from trajectory models are typically based on very limited simulation times, and thus the use of those models to derive a peak 8-hr averaged ozone-reactivity scale is questionable. Finally, the promulgation of the new 8-hr NAAQS for ozone is likely to extend nonattainment into larger geographical regions that include rural as well

as urban and suburban areas (Chameides et al. 1997). Thus far, little
work has been done to assess reactivities at these large, regional scales.
Moreover, ozone chemistry at the regional scale and in rural areas has
generally been found to be NO_x-limited (OTAG 1997), where implementation of VOC emission controls and using a VOC-reactivity scale might
prove to be less effective.

OUTSTANDING TECHNICAL ISSUES IN
QUANTIFYING REACTIVITY

The scientific and policy-making communities have made significant
advances in understanding and implementing methodologies for quantifying VOC ozone-forming potential using the concept of incremental
reactivity. Nevertheless, key issues remain. Among these are the uncertainties in the understanding of the atmospheric chemistry of specific
VOCs, and thus in the ability to quantify their ozone-forming potential,
and the variability in reactivity between different environments. It was
earlier stated that ozone sensitivity to VOC can, in general, vary from
place to place within a given airshed and from episode to episode. Thus,
environmental variability is not limited solely to one city versus another,
but also to different locations within a city and also from one time to
another. Further, it is not apparent that a reactivity scale developed for
high-ozone episodes will be the same as one developed for more typical
conditions. Also, as was found in Los Angeles, the impact on the peak
ozone concentration is not likely to be the same as the impact on ozone
exposure surrogates.

Another important issue relates to the role of NO_x. VOC reactivity,
and its use in control strategies, is of much less relevance in a system and
in locations that are strongly NO_x-limited. Thus, VOC reactivity should
be viewed as a way of providing extra benefits to a strategy based on the
implementation of VOC emissions controls. A major complication can
arise, however, when a given control measure affects NO_x emissions as
well as VOC emissions, especially if the emission changes for the two sets
of precursors are directionally different, which might be the case for
reformulated gasoline using ethanol versus MTBE. Under these circumstances, one can, in principle, derive reactivities for NO_x as well as VOCs
to assess the net impact of the control measure on ozone. However, little

research has been undertaken on the derivation and application of NO_x reactivities. Moreover, as implied earlier, NO_x reactivities would likely be even more dependent upon location and episodic conditions than VOC reactivities. Application of NO_x reactivities for a national ozone mitigation program would therefore be problematic.

Finally, consideration should be given to the future use of reactivity scales for particulate matter (PM) and ozone formation. Similar to ozone, different VOCs can lead to a substantial variation in the formation of secondary particulate matter; many VOCs will form no extra secondary organic particles, but others can lead to a substantial amount. In some cases, the compounds that lead to little ozone formation lead to little PM formation, and those that have a high ozone-forming potential also can form a large amount of particles. In other cases, the opposite is true. Models for simultaneously assessing PM reactivity and ozone reactivity are still under development.

SUMMARY

Ozone atmospheric chemistry involves many thousands of reactions and a similar number of compounds. The two primary precursors to ozone formation are VOCs (and CO) and NO_x, although this, alone, is an oversimplification. There are hundreds of different VOCs emitted into the atmosphere, and no two have the same chemistry; thus, they each have a different impact on ozone. Further complexity comes from the fact that the atmosphere is highly variable, both in its physical and chemical make-up. Thus, not only does ozone formation respond differently to different VOC species, but it will often respond differently to the same compound in different locations or during different episodes at the same location.

A variety of metrics or scales have been proposed to quantify the ozone-forming potential of an individual VOC or a mixture of VOCs arising from a specific source or type of emission. The reactivity paradigm is but one of a number of approaches that have been developed for this purpose. It is based on scientifically sound concepts and can provide a useful approach for policy-makers attempting to decide which VOCs or types of emissions to regulate and to what degree. Indeed, the state of California has already applied the reactivity paradigm to its regulation of motor-vehicle emissions and the fuels used to power those vehicles. Exactly what metric should be chosen is, in part, a question of policy reflecting a set of priorities of the relevant stakeholders.

Within the reactivity paradigm, a number of different scales can be used. Each one provides a measure of the ozone-forming potential of a VOC or mixture of VOCs under a specific set of conditions. In this report, the maximum incremental reactivity (MIR) scale is used as the primary quantitative measure of ozone-forming potential. That scale reflects the ozone-forming potential of VOCs under conditions where ozone control is most sensitive to decreases in VOCs and is also the scale that the state of California has proposed using for its regulatory applications. For simplicity and in the interest of brevity, the term "reactivity" is used to denote the MIR, unless stated otherwise. Moreover, reactivity is expressed in a variety of ways. The specific reactivity, derived from box modeling, is the reactivity normalized to the change of mass of VOC emissions and has units of grams of ozone formed per grams change of VOC emitted or grams of ozone per grams change of VOC. The total reactivity is obtained by multiplying the specific reactivity by the mass of VOC emissions per mile driven and has units of grams of ozone per mile. The relative reactivity is a unitless quantity which is derived by dividing the (specific or total) reactivity of a compound or class of compounds by the (specific or total) reactivity of some reference VOC, standard VOC, or VOC mixture. Sometimes the term absolute reactivity is used in this report to denote either the specific or total reactivity as a way of distinguishing them from the relative reactivity. Each of these terms is listed in Table 3-9.

There are a number of limitations to the reactivity approach that should be borne in mind. Because the ozone-forming potential of VOCs can vary from locale to locale, it should not, in principle, be uniformly applied to the entire nation, except to facilitate regulatory application. Ideally, its use as a certification tool on a nationwide basis would allow for regionally-specific applications and, potentially, the development of regionally-tailored control strategies. Assessing the economic viability of implementing regionally-specific rules for certifying RFGs is beyond the scope of this report.

In its current state of development, a limitation of the use of a reactivity approach beyond full certification is that it only considers the ozone-forming potential of VOCs and CO. Thus it is of less use for

TABLE 3-9 Terms Used in the Report to Denote Reactivity[a]

Term	Definition	Units
Specific reactivity	Reactivity (as MIR) normalized to the change in mass of VOC emissions	g O_3/g change VOC
Total reactivity	Product of specific reactivity (as MIR) and the mass of VOC emissions per mile driven	g O_3/mile
Absolute reactivity	Either the specific or total reactivity	g O_3/g change VOC or g O_3/mile
Relative reactivity	Ratio of the specific or total reactivity (as MIR) of a compound or class of compounds to that of some reference or standard VOC or VOC mixture	Unitless

[a]In this report, the term reactivity is used to denote the maximum incremental reactivity (MIR). MIR reflects the ozone-forming potential of VOCs under conditions that are most sensitive to these VOCs.

developing VOC-based control strategies in areas where only NO_x emission controls are needed. The reactivity approach is also of limited utility in assessing the impacts of control strategies that increase (decrease) VOCs emissions, while decreasing (increasing) NO_x emissions. As it turns out, this might occur in the case of motor-vehicle emissions using specific types of RFG blends.

It is also important to note that the determination of reactivities for VOCs is a computational process that requires the application of a numerical model. The types of models that can be used for this purpose range from rather simplistic trajectory or box models to very complex, three-dimensional grid-based or Eulerian models. All of those models rely on a chemical mechanism for simulating the ozone-forming process, and a variety of algorithms for representing this chemistry have been adopted. Although differences between model results do occur (for example, in the case of the reactivities of the aromatics), in general, the relative reactivity of VOCs derived from different models and models using different chemical mechanisms tend to be reasonably consistent. For this reason, it is believed that the uncertainties (or potential errors) in reactivities can be minimized by focusing on relative as opposed to absolute reactivities.

In general, the 2-σ (or 95% confidence level) uncertainty in the relative reactivities in most of ubiquitous VOCs (that have been studied extensively) is about 20-40%. The relative reactivity of a composite set of VOCs arising from a single source, such as motor vehicles, tends to be somewhat smaller (i.e., about 15-30%). Much of the uncertainty in this later case arises from potential errors in defining the speciation of the emissions as opposed to those associated with the chemistry of the species. For this reason, the use of relative reactivity to assess the ozone-forming potential of different sources is best suited to situations where the reactivity of the emissions is quite different. As will become apparent in later chapters, this tends to not be the case for emissions from motor vehicles using slightly different RFGs. That will, in turn, limit the ability to use reactivity to distinguish robustly between the air-quality benefits of various RFG blends.

4

Motor Vehicles As a Source of Ozone Precursors

THE PRIMARY REGULATED emissions from gasoline-fueled automobiles and trucks—volatile organic compounds (VOCs), nitrogen oxides (NO_x), and carbon monoxide (CO)—all contribute to the formation of ground-level ozone. Moreover, these mobile sources distribute ozone precursors more broadly than stationary sources. This chapter reviews motor-vehicle emissions from light-duty vehicles (LDVs) and, in particular, those vehicles fueled by gasoline (LDGVs). It focuses on the regulation of these emissions and the historical effect they appear to have had on emission inventories and air-quality trends. Deviations of actual emissions from levels set by regulatory intent to control them and the probable reasons for such deviations are then reviewed.

LIGHT-DUTY VEHICULAR EMISSIONS BY SOURCES AND REGULATION

Vehicular-Emissions Sources

Gasoline-fueled automobiles and light trucks (which include certain vans and sport utility vehicles) are important sources of VOCs, NO_x, and CO. VOCs that arise from engine combustion exhaust include many different

73

species, some of which were not present in the original fuel but were created in the combustion reaction and leave the tailpipe without being fully oxidized. Evaporative VOC emissions, on the other hand, result from vapor escaping the fuel storage and transfer system, as well as from fuel leakage, and are thus independent of combustion. NO_x and CO emissions are generated during the combustion process and these only occur in the exhaust.

Tailpipe emissions of VOCs, CO, and NO_x are measured for emissions certification by means of the Federal Test Procedure (FTP), during which a test vehicle is driven on a chassis dynamometer over a prescribed driving schedule. The car is first stored with the engine off ("soaked") at room temperature for at least 12 hr. Then it is started with a cold engine, run over an 18-cycle urban-like driving pattern, stopped for a 10-min hot soak, restarted, and rerun over the first 5 of those 18 cycles. This 18-cycle driving pattern, known as the LA-4 schedule, was developed in the late 1960s to represent a commute to work in the typical Los Angeles traffic of the time. Following some minor modification, it became the basis of the federal U.S. Environmental Protection Agency (EPA)-mandated certification testing procedure for LDVs in 1975 (and is therefore also called the FTP75).

As illustrated in Figure 4-1, the entire LA-4 schedule covers 7.45 miles (mi) at an average speed of 19.6 miles per hour (mph). After adding the 5 repeat cycles following the hot soak, the entire FTP urban driving schedule covers 11.1 miles of driving in 31 min, excluding the 10-min hot soak.

During the FTP, tailpipe exhaust is collected in three bags: the so-called cold bag for the first 5 cycles of driving, the stabilized bag for the next 13, and the hot bag for the 5 repeat cycles following the hot soak. For regulatory purposes, the measured mass emissions from each bag are substituted in a prescribed equation to determine the emission rate per unit of travel (in this case, grams per mile) of each regulated emission.

Evaporative emissions, including those resulting from leaks of liquid fuel, are measured separately using a variable-temperature SHED (sealed-housing-for-evaporative-determination) facility; i.e., an instrumented temperature-controlled room in which the test vehicle is housed. The fuel system of each car includes an evaporative canister containing a bed of activated carbon particles that adsorb most of the fuel vapor that might otherwise escape to the environment. The canister is connected to both the fuel-tank headspace and the engine intake. During

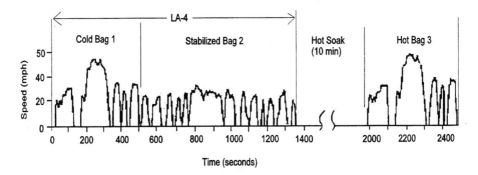

FIGURE 4-1 The Federal Test Procedure urban driving schedule covers 11.1 miles of driving in 31 min, excluding the 10-min hot soak. Source: Adapted from Davis 1998.

normal engine operation, stored VOC vapor is purged from the canister, drawn into the engine by intake-manifold vacuum, and consumed in combustion. However, the system is not 100% effective. Escape routes for evaporative gases include the engine intake and vents in the fuel tank as well as the canister itself.

Evaporative emissions, including those resulting from leaks of liquid fuel, can be classified into five categories: diurnal, hot soak, running loss, resting loss, and refueling loss.[1] The characteristics and causes for each of these emissions are described briefly below. Originally, only diurnal evaporative emissions were regulated. In more recent years, hot-soak emissions were added to the diurnal emissions for regulatory purposes, with a separate limit being placed on the running loss. The first refueling-loss standard began with a 3-year phase-in period on passenger cars in 1998. These emissions are controlled by an on-board refueling vapor canister.

Diurnal emissions occur because the fuel tank of a parked car "inhales" air at night as the tank cools, then "exhales" a mixture of air and fuel vapor during the day as tank temperature rises. Diurnal emissions

---

[1]In addition to evaporative emissions, non-combustion emissions from motor vehicles can arise from dripping and leaking of fuel. The term "nonexhaust emissions" is used to denote the total of evaporative emissions and the emissions that arise from fuel leakage.

tend to increase linearly with available tank headspace, and are also very sensitive to tank temperatures and fuel volatility.

Hot-soak emissions occur after vehicle operation has been terminated. These emissions are measured over a 1-hr period after the vehicle has completed a prescribed driving schedule. Hot-soak emissions from a given vehicle depend on the previous driving schedule, ambient conditions, and fuel volatility.

Running-loss emissions occur as the tank is heated during vehicular operation and can result from the following:

- Inefficiency of the in-tank fuel pump and motor.
- Recirculation to the tank (in some vehicles) of excess fuel supplied to the port fuel injectors.
- External heat from the nearby exhaust system.
- External heat from the air flowing under the car from the engine compartment.

Vapor generated during vehicle operation is directed to the canister for transfer to the engine, where it is consumed. The line from the canister to the engine contains a valve that regulates that purge flow. However, when the quantity of vapor, thus generated, exceeds the ability of the engine to consume it and the canister has reached its storage capacity, the excess vapor escapes through the canister vent as "breakthrough" emissions and constitutes the running loss. Thus running-loss emissions depend on the driving schedule, ambient conditions, fuel volatility, type of vehicle, and condition of the control system.

Resting-loss emissions include escape of fuel vapor by means of permeation of nonmetallic components of the fuel system while the vehicle is inoperative. Resting-loss emissions depend on fuel characteristics and design features of the fuel system.

Refueling emissions consist of the fuel vapor displaced from the tank headspace by the new liquid fuel being pumped into the tank. Typically, these vapors are stored in the same canister used to control the other categories of evaporative emissions. Refueling emissions occur when these vapors escape, and depend on the volume of fuel pumped and on the respective temperatures and compositions of the fuel remaining in the tank and the pumped fuel.

Control Standards for These Sources

Since their inception, emissions standards have been progressively tightened. The trend toward greater stringency in tailpipe and diurnal evaporative controls through 1993 is tabulated separately for federal and California standards in Table 4-1. These standards were to be satisfied through 50,000 miles of driving; consequently, to ensure compliance, manufacturers calibrated new car emissions levels to be substantially below the specified 50,000-mile level.

Standards in effect from 1993 onward are listed in Table 4-2. Standards are defined for both 50,000- and 100,000-mile compliance. The standards for high mileage accrual are intended to preclude excessive emissions as accumulated service surpasses 50,000 miles. As shown, California has defined a family of low-emissions vehicles: the transitional low-emissions vehicle (TLEV), the low-emissions vehicle (LEV), and the ultra-low-emissions vehicle (ULEV). Not shown is the zero-emissions vehicle (ZEV), which has no tailpipe emissions. The only vehicle currently qualifying as a ZEV is a dedicated electric car or light truck. (However, even in this case, use of the vehicle does in fact result in ozone precursor emissions whenever fossil fuels are burned to generate the electricity required for battery charging.) California allows manufacturers to mix conventional vehicles and members of the low-emissions family, within certain constraints, in a manner that forces a gradual reduction in fleet-average emissions in successive years.

At 50,000 miles, the federal Tier I standards now in place entail reductions in combined tailpipe and crankcase emissions, from the average precontrol car, of 98%, 96%, and 90% for NMHC, CO, and NO_x, respectively. For the California ULEV, these reductions increase, respectively, to 99+%, 98%, and 95%.

Major manufacturers have volunteered to build cars to national low-emissions vehicle (NLEV) specifications having NMHC, CO and NO_x standards equal to those of the California LEV in Table 4-2, making them available in the Northeast in 1999 and nationwide in 2001. The 70% reduction in tailpipe NMHC and the 50% reduction in NO_x with the NLEV, compared with Tier 1 vehicles, pursue a national improvement in air quality earlier than had been anticipated by regulatory schedules. In addition, manufacturers are moving voluntarily to produce light trucks

TABLE 4-1 Emissions Standards for Automobiles (allowable emission levels up through 50,000 miles of driving)

Model year	Federal HC (g/mi)	Federal CO (g/mi)	Federal NO$_x$ (g/mi)	Federal PM[a] (g/test)	Federal Evap (g/mi)	California HC (g/mi)	California NMHC[a] (g/mi)	California CO (g/mi)	California NO$_x$ (g/mi)	California PM[a] (g/mi)	California Evap (g/test)	California HCHO[a] (g/mi)
Pre-control	14.7[b]	84.0	4.1		47	14.7[b]		84.0	4.1		47	
1966	6.3	51.0	(6.0)[c]			6.3		51.0	(6.0)[c]			
1968	6.3	51.0				6.3		51.0				
1970	4.1	34.0				4.1		34.0			6	
1971	4.1	34.0				4.1		34.0	4.0		6	
1972	3.0	28.0				2.9		34.0	3.0		2	
1973	3.0	28.0	3.0			2.9		34.0	3.0		2	
1974	3.0	28.0	3.0			2.9		34.0	2.0		2	
1975	1.5	15.0	3.1[d]		2	0.9		9.0	2.0		2	
1977	1.5	15.0	2.0		2	0.41		9.0	1.5		2	
1978	1.5	15.0	2.0		6[d]	0.41		9.0	1.5		6[d]	
1980	0.41	7.0	2.0		6		0.39	9.0	1.0		2	
1981	0.41	3.4	1.0		2		0.39	7.0	0.7		2	
1982	0.41	3.4	1.0	0.60	2		0.39	7.0	0.7		2	
1983	0.41	3.4	1.0	0.60	2		0.39	7.0	0.4		2	
1984	0.41	3.4	1.0	0.60	2		0.39	7.0	0.4	0.60	2	
1985	0.41	3.4	1.0	0.60	2		0.39	7.0	0.4	0.40	2	

1986	0.41	3.4	1.0	0.60	2		0.39	7.0	0.4	0.20	2	
1987	0.41	3.4	1.0	0.20	2		0.39	7.0	0.4	0.20	2	
1989	0.41	3.4	1.0	0.20	2		0.39	7.0	0.4	0.08	2	
1993	0.41	3.4	1.0	0.20	2		0.39	7.0	0.4	0.08	2	0.015

[a]Particulate matter, applicable to diesel cars only; NMHC = nonmethane hydrocarbons; HCHO = formaldehyde.
[b]Includes 4.1 g/mi of crankcase emissions, fully controlled by 1966.
[c]Uncontrolled NO_x increased as HC and NO_x standards were implemented.
[d]Change in test procedure.
NOTE: Empty cells indicate no standards in place for those years.
Source: Adapted from Calvert et al. 1993.

80 OZONE-FORMING POTENTIAL OF REFORMULATED GASOLINE

and vans, rated above the maximum weight range covered by the federal standards of Table 4-2, in a manner that will enable these vehicles to be certified to the more stringent passenger-car standards. This should have a substantial air-quality benefit because these heavier vehicles accounted for nearly half of new-vehicle sales to the general public in 1998.

MAGNITUDES AND TRENDS OF LIGHT-DUTY VEHICULAR EMISSIONS

To gauge the potential air-quality benefit from the use of reformulated gasolines, in general, and specific oxygenates within these gasolines, it is useful to review the magnitudes and trends in the emissions of VOC, CO, and NO_x from motor vehicles and other sources. The 55-year trends illustrated in Figures 4-2, 4-3, and 4-4 for VOC, NO_x, and CO, respec-

TABLE 4-2 Automobile Emissions Standards, Tier I and Beyond (standards for 50,000 miles or 5 years (100,000 miles or 10 years))

	NMHC (g/mi)	NMOG[a] (g/mi)	CO (g/mi)	NO_x (g/mi)	HCHO (g/mi)
Federal					
Tier I (1994)	0.25 (0.31)		3.4 (4.2)	0.4 (0.6)	
Tier II (2003)	(0.125)		(1.7)	(0.2)	
California					
Conventional vehicles (1993)[b]		0.25 (0.31)	3.4 (4.2)	0.4 (0.6)	0.015 (0.018)
TLEVs (starting in 1994)[b]		0.125 (0.156)	3.4 (4.2)	0.2 (0.3)	0.015 (0.018)
LEVs (starting in 1997)[b]		0.075 (0.09)	3.4 (4.2)	0.2 (0.3)	0.015 (0.018)
ULEVs (starting in 1997)[b]		0.04 (0.055)	1.7 (2.1)	0.2 (0.3)	0.008 (0.011)

[a]NMOG = nonmethane organic gases (NMHC + oxygenated HC).
[b]Measured NMOG adjusted for reactivity, relative to conventional gasoline.
Source: Adapted from Calvert et al. 1993.

tively, are based on inventory estimates compiled by EPA. In viewing these inventories, it should be noted that large uncertainties are typically associated with emission inventories, especially those arising from motor vehicles. Historically, it has been found that the contribution of emissions from mobile sources had been underestimated, and each time new information and data became available, these emissions had to be revised upward (NRC 1991). Moreover, the accuracy of contemporary mobile source emission inventories remains the subject of some debate (Sawyer et al. 1998).

From Figure 4-2, it can be seen that the contribution to anthropogenic VOCs from highway vehicles appears to have peaked around 1970. By 1995, this share had declined to 28% of the anthropogenic total, by which time industrial processes were estimated to account for 47% of

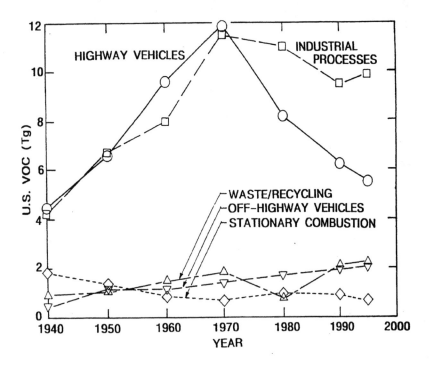

FIGURE 4-2 Estimated trends in VOC emissions from various types of sources in the United States. Emissions are presented in units of teragrams (Tg). 1 Tg = 10^6 metric tons. The contribution from "Highway Vehicles" includes LDVs, the automobiles and light trucks that are the subject of this study, and heavy-duty vehicles (HDVs), larger trucks and buses. Source: Adapted from Davis 1997.

anthropogenic VOCs. The remainder was attributable primarily to waste disposal and recycling (5%), off-highway vehicles (12%), and stationary fuel combustion (5%).

In the case of NO_x emissions, Figure 4-3 indicates that highway vehicles accounted for about 31% of all anthropogenic NO_x in 1995. Approximately 70% of this came from LDVs, which are powered primarily by gasoline engines, with the balance produced by HDVs, which are primarily diesel-powered. In 1995, stationary combustion accounted for 45% of anthropogenic NO_x, 20% came from nonhighway vehicles, 3% was attributable to industrial processes, and the balance came from miscellaneous sources.

Figure 4-4 indicates that highway vehicles have long dominated national CO sources. In 1995, they were responsible for 60% of the CO total. Adding in off-highway (off-road) transportation sources, the entire transportation sector was responsible for about 80% of the national CO emissions that year.

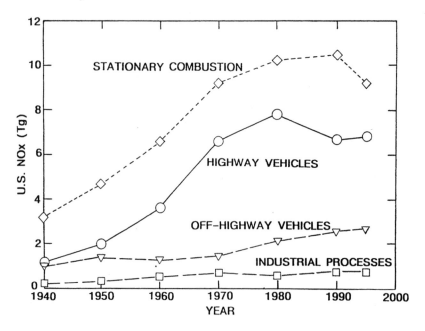

FIGURE 4-3 Estimated trends in NOx emissions from various types of sources in the United States. Emissions are presented in units of teragrams (Tg). 1 Tg = 10^6 metric tons. The contribution from "Highway Vehicles" includes LDVs, the automobiles and light trucks that are the subject of this study, and heavy-duty vehicles (HDVs), larger trucks and buses. Source: Adapted from Davis 1997.

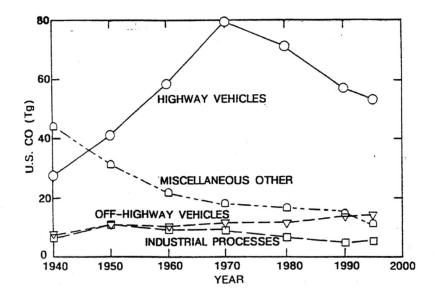

FIGURE 4-4 Estimated trends in carbon monoxide (CO) emissions from various types of sources in the United States. Emissions are presented in units of teragrams (Tg). 1 Tg = 10^6 metric tons. The contribution from "Highway Vehicles" includes LDVs, the automobiles and light trucks that are the subject of this study, and heavy-duty vehicles (HDVs), larger trucks and buses. Source: Adapted from Davis 1997.

Comparison of the national inventory estimates with inventory estimates for California developed by the California Air Resources Board (CARB) reveals some substantial differences. For example, CARB estimates that, in 1995, 25% of its statewide emissions of VOC (total organic gas emissions), 44% of its reactive organic gas (ROG) emissions, and 60% of its NO_x emissions were from on-road vehicles (http://www.arb.ca.gov/ceidars/emssumcat.submit_form). (VOCs include nonreactive organic compounds that are not included within the ROG category and hence emissions of VOCs are greater than those of ROGs.) In contrast, the federal inventory for 1995 ascribes 28% of VOC emissions and 31% of NO_x emissions to on-road vehicles (Davis 1998). The differences in the on-road-vehicle contribution (especially for NO_x) in the two inventories are most likely indicative of the unique characteristics of California as compared to the rest of the nation. However, the possibility that they also arise, at least in part, from inaccuracies in one or both inventories cannot be ruled out.

Given the historical trends in ozone precursor emissions, it is interesting also to review what the corresponding achievements have been in ozone reduction. National air quality and emissions have been reviewed for the period from 1986 to 1995 (EPA 1996a). In Figure 4-5A, the arithmetic mean ozone concentration from 573 measuring sites (normalized using the mean for 1986) is plotted versus year. This trend is compared with the normalized trends in national emissions of VOCs, NO_x, and CO over the same period in Figures 4-5B, 4-5C, and 4-5D, respectively. The substantial year-to-year variability in the ozone trace reflects annual variations in meteorology. For example, before 1996, 1988 was the third hottest year of the century, and temperatures were also high in 1995. Because high ambient temperature is conducive to ozone formation, it is not surprising that ozone levels were, on average, higher in those years. EPA endeavored to correct these annual data for variations in meteorology and concluded that average ozone is decreasing about 1% per year. Additional discussion of variability in ozone trends is contained in Chapter 6 of this report. Over the same period, national anthropogenic emissions of VOCs, NO_x, and CO are estimated to have decreased by about 9%, 2% and 16%, respectively, with emissions attributed to on-road vehicles decreasing by about 31%, 2%, and 20%, respectively. Over this same period, vehicle miles traveled (VMT) by on-road vehicles increased 32%, so that the actual reductions in vehicular emissions of all precursors in units of g/mi were quite substantial (as would be expected from the data in Table 4-1).

Relating the national trend in ozone to the trends in precursor emissions is problematic because of such important influences as variations in precursor reactivity, nonuniformity in the geographical distribution of precursors, and meteorological effects. Nevertheless, the data clearly reflect improvement in air quality over the decade that is likely attributable at least in part to: (1) the advent of more stringent standards for vehicles that gradually replace old vehicles built to more lenient emissions standards than current models; (2) maturation of new-vehicle emissions control hardware and software as field experience accumulated; and (3) recent improvements in gasoline properties. The air-quality improvement is reported despite the increase in VMT, and an increasing preference among consumers for light trucks and vans that emit more precursors.

It is also apparent that during this period, the contribution of LDVs to ground-level ozone pollution has decreased substantially. According to

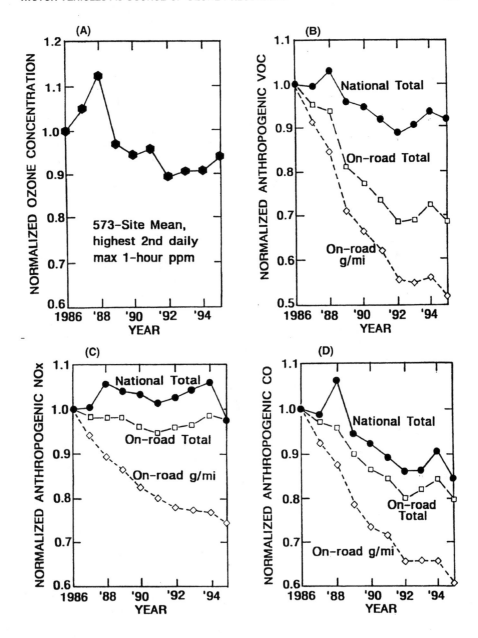

FIGURE 4-5 Normalized trends in the arithmatic mean ozone concentration (A), VOC emissions (B), NO$_x$ emissions (C), and CO emissions (D) over the same period. Source: Adapted from EPA 1996a.

EPA, the contribution of on-road vehicles to VOC and NO_x emissions has decreased by about 30% and 3%, respectively. One consequence of this decreasing contribution is that ozone mitigation strategies based on further reductions in motor-vehicle emissions such as that of the reformulated gasoline programs must necessarily also have a reduced potential to improve air quality. For example, if ozone concentrations responded linearly to a reduction in VOC concentrations, a reduction of 20% in VOC emissions from a reformulated gasoline program might decrease ozone by only about 6%, given that on-road vehicles are currently responsible for about 30% of the total emissions. Even if the contribution of VOC emissions has been underestimated by a factor of 2, the ozone reduction would only be a little more than 10%. In reality, the ozone reduction would be significantly smaller since the response of ozone concentrations to VOC reductions are generally less than linear. As discussed in more detail in Chapter 6, this shrinking contribution to ozone precursors from gasoline-powered motor vehicles makes it very difficult to discern the impact of reformulated gasoline on ambient ozone concentrations, let alone distinguish between the effects of different reformulated gasoline blends. This, however, should not be interpreted to mean that emissions controls on LDVs are not important. Clearly they are; it is just that discerning incremental benefits becomes increasingly difficult as the relative contribution of LDV emissions decreases.

It is also relevant to note that the contribution from motor vehicles might very likely continue to shrink. For example, Figure 4-6 illustrates the estimated and projected emissions for VOCs (shown as TOGs) and NO_x in Los Angeles, New York, and the Chicago-Milwaukee region for 1988, 2000, and 2010 by the Auto/Oil study (AQIRP 1997a). In these projections, use of conventional gasoline (representing a 1988 national average composition) was assumed for the base year and use of various reformulated gasoline blends were assumed for future years. "Other manmade" sources include diverse sources such as power lawn mowers, earth-moving equipment, surface coatings, and solvents and cooking and baking activities. Substantial reductions in LDV emissions were projected for each city. From the base year to 2010, decreases in the LDV emissions were estimated to be from 74% to 92% for VOCs and 54% to 69% for NO_x, depending on the city. At the time when this modeling was performed, the large reduction estimated for LDV emissions anticipated benefits from replacement of older vehicles, lower vehicle emission standards, on-board diagnostics, reduced gasoline vapor pressure, refor-

87

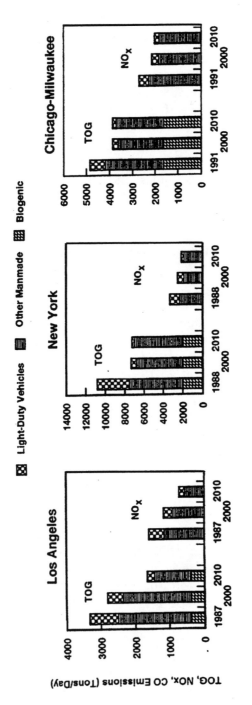

FIGURE 4-6 Estimated emissions in base years and future years in Los Angeles, New York City, and the Chicago-Milwaukee region. Source: Adapted from AQIRP 1997a.

mulated gasoline, and more stringent inspections and maintenance programs. CARB has also projected large reductions in emissions from on-road mobile sources (which include HDVs). For example, these sources are expected to lower their share of statewide ROG emissions from 44% in 1995 to 18% in 2010 (http://www.arb.ca.gov/emisinv/ emsmain/emsmain.htm). If those projections turn out to be accurate, the probable impact from subtle changes in RFG blends will be further reduced. However, it should be noted in this regard that projections of future mobile source emissions depend upon assumptions concerning trends in technology, economics, and human behavior and, as a result, are highly uncertain.

INFLUENCE OF DRIVING PATTERNS ON EMISSIONS VARIABILITY

As discussed above, regulation of LDV exhaust emissions is based on the FTP which is, in turn, built around the LA-4 driving schedule. There is, however, growing concern that this driving schedule is not able to characterize accurately emissions from LDVs under normal driving conditions (Darlington et al. 1992; Kelly and Groblicki 1992). Three potentially important sources of error are discussed below: off-cycle transient events, underrepresented events, and variable events.

Off-Cycle Transient Events

When the LA-4 driving schedule was first devised, the ability of existing chassis dynamometers to accommodate high vehicular accelerations was limited. Consequently, accelerations on the schedule were arbitrarily restricted to a maximum of 3.3 mph/sec. However, data from instrumented cars in typical traffic have shown peak accelerations as high as 15 mph/sec at 20 mph, with somewhat lower rates at higher speeds but all in excess of 3.3 mph/sec (Ross et al. 1995). This raises the possibility that the FTP misses important aspects of typical driving that could result in undetected, increased off-cycle emissions of VOCs, CO, and NO_x (see Text Box 4-1). On the other hand, analysis of in-use survey data indicates that driving at an air-to-fuel ratio of 12% richer than stoichiometric occurs only about 1% to 2% of the time (Ross et al. 1995),

TEXT BOX 4-1 Power Enrichment Can Affect Exhaust Emissions

Warmed-up, conventional gasoline engines are now calibrated to operate under most circumstances at or very near the stoichiometric air-to-fuel ratio to accommodate the three-way catalytic converter. At this ratio (about 14.7:1 for gasoline), a fuel is combusted nearly completely with almost no air appearing unutilized in the combustion products. However, as the throttle is gradually opened to provide more than about 75% of the maximum power available at any given speed, the mixture is gradually enriched from the stoichiometric to a lower air-to-fuel ratio for a number of reasons. First, theoretically, the engine is capable of producing about 5% more power when the mixture is enriched about 10% beyond the stoichiometric ratio. As vehicular performance potential is determined by maximum power, this degree of enrichment allows about a 5% reduction in piston displacement for the same performance potential, making it possible to meet performance criteria with a smaller engine with better fuel economy. Second, for a given air-to-fuel ratio, the exhaust gas temperature is highest at full throttle. When the average full-throttle mixture is set at the stoichiometric ratio, there is always a small amount of oxygen in the exhaust stream, ideal chemistry notwithstanding. Should one cylinder experience an instance of poor combustion, raw fuel, oxygen, and a high exhaust temperature can co-exist in the catalytic converter. This invites a significant upward excursion in catalyst temperature that can hasten deterioration, and, in severe cases, even lead to destruction of the catalyst. To prevent this from occurring, the mixture is enriched. Third, to suppress combustion knock (see Chapter 5), some engines are calibrated even richer than the maximum-power ratio at full throttle. This allows use of a higher compression ratio for better fuel economy at part load, at which the engine operates most of the time.

With respect to emissions, as an engine approaches full throttle, power enrichment significantly decreases engine-out NO_x. However, vehicular tests have shown that the catalytic converter can pass an increasing fraction of this raw NO_x as the mixture is enriched (Ross et al. 1995). This catalyst behavior might be the result of insufficient exhaust residence time in the converter at high engine-flow rates. Power enrichment at high loads also substantially increases engine-out CO and VOCs just as the rich mixture is depriving the catalytic converter of enough oxygen to destroy those emissions.

suggesting that power enrichment might not make a major contribution to LDV emissions resulting from typical driving. Nevertheless, the Clean Air Act Amendments of 1990 directed EPA to devise a Supplemental FTP (SFTP) to assess some of the real-world emissions not properly addressed in the LA-4 schedule. It is intended to add a more aggressive driving pattern, a change made possible by improvements to chassis dynamometers since the 1960s.

As defined by the EPA rule of October 1996 (EPA 1996b), beginning with vehicular certifications for model year 2000 (MY2000) and phasing to 100% application for MY2002 (MY2004 for larger light trucks), driving schedule US06, illustrated in Figure 4-7, will be run immediately following the current FTP to contribute exhaust to a fourth collection bag. Emissions from this bag will be weighted in a calculation together with those from the first three to arrive at the exhaust certification emission rate (grams per mile). An improved correction test for air-conditioning load, called the SC03, will also be conducted. The SFTP is expected to lead to more realistic control of emissions in real-world driving.

Underrepresented Events

The warm-up time for emissions-control components, especially for catalysts, varies across makes and models of vehicles. Because those components do not operate at peak efficiency during warm-up, emissions can be unusually high for a brief period every time a vehicle is cold-started or restarted. Other factors remaining equal, the magnitude and duration of this excursion in a given vehicle depend on ambient temperature and the length of time the engine has been shut down. Multipurpose, chained trips involving at least one intermediate stop of 15 min or less, such as from home to a series of retail establishments, to school, and to a final destination, are now recognized as becoming increasingly common, accounting for perhaps as many as half of the total trips (FHWA 1997). For this reason, the extra emissions associated with a chain of short trips are a growing concern.

The current FTP includes one cold start and, after a 10-min soak with the engine off, one hot restart over 11.1 miles. It does not directly incorporate any hot soaks, but those soaks can be an important contributor to total emissions in chain driving. For regulatory purposes, hot-soak

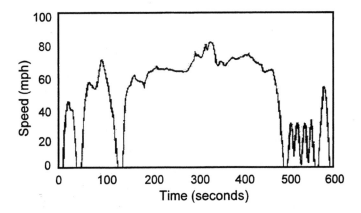

FIGURE 4-7 Driving schedule US06 test procedure to be used immediately after the Federal Test Procedure to contribute exhaust to a fourth collection bag. Source: EPA 1996b.

emissions are currently measured separately as part of SHED testing (for which the standard is based on a maximum total mass of evaporative VOCs) and thus are not integrated with tailpipe emissions. Hot soaks and other important aspects of modern driving practices might currently be misrepresented in the overall certification procedure.

Other sources of evaporative emissions are affected by in-use driving activities. The major source of unburned hydrocarbons in the fuel storage and supply system of a fuel-injected vehicle is the fuel-tank vapor space. The rate at which tank vapors are collected and purged by the evaporative emissions-control canister (as part of the running losses) tends to be a function of both driving pattern and engine-on time. For example, high purge rates are generally associated with high constant-speed driving, whereas low rates tend to accompany lower-speed driving with more frequent stops (Kishan et al. 1993).

A departure from FTP conditions that occurs in real driving is the existence of highway slopes, or grades. A U.S. Department of Transportation survey concluded that 10% of nationwide driving occurred up grades of 0.5% to 1.0%, 12% up 1% to 3% grades, 7% up 3% to 5% grades, and 3% up grades of over 5% (EPA 1980). Driving uphill increases the power requirement of the vehicle by an amount proportional to the grade and the speed of the vehicle. Depending on engine displacement and calibration, the need for extra power can lead to fuel enrichment. In general, the lower the power-to-weight ratio of the vehicle, the more likely it is to encounter power enrichment on upgrades. It is improper to focus exclusively on the increased emissions that accompany increased fuel flow when traveling uphill. A vehicle driven uphill must eventually travel downhill. Although the mass emissions may be lowered as the fuel flow is decreased on a downgrade, the extra emissions that were produced on the upgrade are not exactly canceled by an equivalent reduction in emissions on the downgrade.

Variable Events

A factor of growing significance that affects emissions is increasing urban traffic congestion. During peak hours in some cities, expressways built for high speed resemble parking lots full of vehicles with all their engines idling. Because of the low fuel rate during idling, it has been concluded that excess emissions from idling due to traffic congestion are relatively low in a properly functioning vehicle (Ross et al. 1995). On the other hand, the phenomenon frequently observed on urban freeways of alternating between hard acceleration as congestion diminishes and braking to a stop at the back of a queue results·in more fuel consumption and emissions than if the same distance were covered at a constant speed. Improved traffic management schemes, including variable message signing and increased highway automation, are expected eventually to improve this situation by smoothing traffic flow, but the implementation of such systems will be gradual. Meanwhile, the differences in power demand and the random distribution of individual vehicles operating in these "sawtooth" driving conditions render estimation of their aggregate emissions very uncertain.

Irrespective of the habits of individual drivers, highway conditions collectively give rise to important departures from driving norms. Recurrent congestion results in trips of longer duration and, concomitantly, more aggregate engine-on time than would be the case in the absence of excessive traffic. This in turn can lead, for example, to higher fuel-tank heating, with increases in the temperature of the fuel itself. Even in properly functioning vehicles, this might contribute to high running-loss emissions (Kishan et al. 1993). In late-model vehicles, the tank-heating problem is being addressed by eliminating the practice of recirculating hot, unused fuel from the injectors back to the tank.

Another variable environmental factor affecting power requirement is wind. The power expended in overcoming aerodynamic drag increases as the cube of wind speed relative to the vehicle. A car driving 20 mph into a 20-mph head wind encounters 8 times the aerodynamic drag of that same car driving 20 mph in still air. As with grades, however, there is somewhat of a compensating effect when that same car has a 20-mph tail wind. In most urban driving, vehicular speed is low enough that the effect of wind on demand for total vehicular power, hence on fuel consumption and emissions, is much less than would be the case in highway driving.

Even in the absence of wind, the aerodynamic efficiency and performance of a vehicle can be negatively affected by loads such as a rooftop carrier, which adds drag as well as mass. Pulling a trailer adds rolling resistance, weight, and drag.

Yet another real-world consideration is the use of air-conditioning. On the present FTP, increasing the dynamometer load 10% simulates the use of an air conditioner. That this simple expedient cannot accurately reflect the influence of the air conditioner on power requirement, hence emissions, is obvious from the fact that when the car is stationary with the engine idling, 10% of the dynamometer load is zero. In a real car the air conditioner is extracting more than zero power from the engine while the vehicle is stationary. The SFTP air-conditioner load cycle is intended to correct this discrepancy.

EMISSIONS DETERIORATION AND PROSPECTS FOR DETECTION

The engines propelling recently manufactured cars incorporate technologies unheard of when emissions regulation began. Electronically controlled port fuel injectors have replaced the carburetor of yesterday. An on-board computer has been incorporated into a closed-loop control system that oscillates the air-to-fuel ratio within a narrow window about the stoichiometric ratio. This ensures that a three-way catalyst maintains a high conversion efficiency for VOCs, CO and NO_x concurrently, as illustrated in Figure 4-8. Typically, the on-board computer closes the mixture-ratio control loop using signals from an inlet airflow sensor, the fuel injectors, and an exhaust-gas oxygen sensor that indicates whether the air-to-fuel ratio is being maintained within the desired range. In addition, exhaust-gas recirculation (EGR) is employed to decrease the flame temperature for lower NO_x emission.

During the cold start that initiates the FTP, the room-temperature

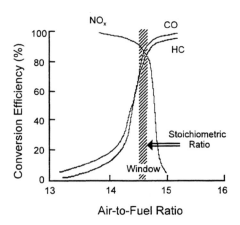

FIGURE 4-8 Illustration of the range of control efficiency for a three-way catalyst with respect to NO$_x$, CO, and VOC emissions. Source: Adapted from Canale et al. 1978.

catalyst is ineffective, as illustrated in Figure 4-9. Similarly, the exhaust-gas oxygen sensor must undergo a warm-up period before it becomes functional, although that problem is now typically minimized by electrically heating the sensor during starting.

Unfortunately, the ineffectiveness of the cold catalyst coincides with the need for a rich air-to-fuel ratio to ensure engine starting, because the spark plug cannot ignite the air-to-fuel mixture unless the ratio of air to fuel vapor is within the flammability limits of the fuel. Because gasoline is a mixture of hydrocarbons with a range of volatilities and the low-volatility components do not vaporize in the cold-cylinder environment, extra fuel is needed to increase the quantity of high-volatility components present to ensure a combustible air-to-fuel ratio for starting.

A consequence of the simultaneous ineffectiveness of a catalytic converter not yet heated to its operating temperature and the need for a temporarily rich starting mixture is that, in recent low-emissions cars, as much as 80% of the FTP VOCs are emitted during the first 1 or 2 min after the cold start. Also, because of the high conversion efficiency of the warmed-up catalyst, tailpipe emissions are profoundly affected by excessive deterioration of catalyst efficiency as mileage is accumulated. A manufacturer makes allowance for reasonable deterioration by setting emissions performance targets for new cars at a level well below (more stringent than) the 50,000-mile standards.

Given the high conversion efficiency of the contemporary warmed-up

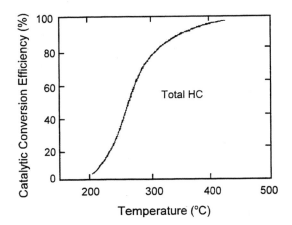

FIGURE 4-9 Relationships between catalytic conversion efficiency and catalyst temperature. At ambient temperature, the catalytic converter is ineffective. Source:
Adapted from Heywood 1988.

catalyst, there is strong motivation to decrease emissions during the cold
start by shortening catalyst warm-up time. Conserving heat by insulating
the pipe or pipes connecting the engine to the catalytic converter is now
common practice. Electrically heated catalysts have been tried, but their
acceptance is hampered by concern about the increased drain on the car
battery, an issue particularly worrisome in cold northern winter climates.
A more popular trend is to use a second, or warm-up, converter located
close to the exhaust manifold. This minimizes heat loss upstream of the
catalyst. An additional converter might also be placed downstream of
this warm-up converter.

   Although new cars are designed to meet applicable emissions standards over their useful life, there is a continuing need to verify that
vehicles in the hands of the public are satisfying this objective. Many
tests conducted to monitor the emissions from such vehicles indicate that
an unsatisfactorily high proportion of them fail to meet expectations in
use (Calvert et al. 1993).

   On new vehicles, EPA and CARB use a selective enforcement audit to
spot check emissions performance of manufactured vehicles at the end
of the assembly line. Though manufacturers might perform voluntary
quality assurance checks, provisions of 40 CFR 86.603-88(e) statutorily
limit EPA to auditing no more than one in every 300,000 of each manufacturer's model-year production destined for the U.S. market, with

thresholds in the audit count set at 150,000 units. Therefore, at a production level of 450,000 units, there is a transition from one to two audits; at 750,000 units, from two to three audits, and so forth. For those manufacturers producing fewer than 150,000 vehicles for the U.S. market, the annual audit limit is one.

However, if there is evidence of noncompliance with standards, the EPA administrator can issue additional test orders. If the emissions failure rate of new vehicles coming off an assembly line is excessive, the line can be shut down until the problem is fixed. The manufacturer also must recall and repair any such vehicles already produced.

Vehicles already in use are subject to an after-market, in-use test. Each year, EPA and CARB select some engine families and ask a number of owners of vehicles with those engines to submit their vehicles for testing. These vehicles typically have accumulated 30,000 to 50,000 miles of customer service. An excessive failure rate can trigger a recall. This approach has been criticized because it is voluntary. The regulatory agencies cannot force an invited owner to participate, thus imposing an unintended bias on the test sample. Also, from the sample recruited, only properly maintained vehicles are tested.

Inspection and Maintenance (I&M) programs have been instituted in states required to comply with mobile-source air-quality attainment provisions of the Clean Air Act. The object of such programs is to identify LDVs that are significantly out of compliance and have them repaired as a requirement for continued licensing. Previously, this technique was able to identify two common causes of malfunction in control systems: misfueling and tampering. Misfueling a catalyst-equipped car with leaded gasoline led to poisoning of the catalytic converter. With the removal of leaded gasoline from the market (see Chapter 5), this problem has been eliminated. In the 1970s, drivers sometimes tampered with their vehicles by rendering parts of their emissions-control systems nonfunctional in the belief that it would improve driveability or fuel economy. Disconnecting a hose to deactivate the EGR system (an NO_x control technique) was an example. The modern control system is so complex and sophisticated that tampering has become a rarity, and can often be self-defeating with respect to fuel economy or performance.

I&M can, in principle, detect a malfunctioning control system. In practice, however, the test has been too simplified in most locations to detect more than a few possible malfunctions. For example, a frequently employed I&M technique involves only measurement of the CO and VOC concentrations in the tailpipe of a warmed-up idling engine and possibly a visual check for tampering. Recent enhancements to I&M procedures, such as a 240-sec test called the IM240 that mimics the driving loads of key portions of the FTP, involve operating the engine under load on a chassis dynamometer. Although not yet uniformly adopted, IM240 procedures are now employed in areas such as parts of Indiana, Arizona, and Colorado.

Studies of on-road emissions performance have been conducted using a remote sensing technique that involves measuring the absorption of infrared light beamed across a single traffic lane behind a passing vehicle in normal traffic (Stephens et al. 1997). EPA has granted incremental emissions reduction credits to I&M programs that also incorporate such remote sensing. These measurements suggest that despite all precautions, a significant number of high polluters exist in the fleet. However, caution should be exercised in applying this technique (Ross et al. 1995). A single drive-by measurement provides only a snapshot of the tailpipe emission. If it measures emissions concentration before the catalyst has warmed up, or during a heavy acceleration, or during coasting with the throttle closed, measured results correlate poorly with a dynamometer test, which, in this case, would provide a more-comprehensive reflection of actual driving conditions. Correlation can be improved by using a multipass average.

A number of other studies of in-use emissions have been conducted on an irregular basis, including repetitions of new vehicle certification tests on LDVs that have accumulated mileage in customer service. In one such study, it was determined that 80% of the stabilized and hot-soak VOCs came from the worst 20% of the vehicles on the road (AQIRP 1997b). In another series of tests on 1979 to1989 models, 62% of the total CO from the test fleet came from only 7.6% of the vehicles (Ross et al. 1995). To illustrate the significance of these high emitters, if this 7.6% of the fleet were replaced with a like number of cars having the average CO emissions of the rest of the fleet, the total CO from the fleet would be decreased by about 60%. Either repairing high emitters or removing them from the fleet might be as effective as tightening emissions standards (Calvert et al. 1993).

As tailpipe-emissions standards become more stringent, evaporative emissions assume increasing importance. Deterioration of evaporative-emissions controls in the field has received less study than deterioration of exhaust emissions controls. In one examination of approximately 300 in-service vehicles, 15% were judged to have high evaporative emissions (Brooks et al. 1995). Among the high evaporative emitters, problems were found with failed gas caps and caps that were either not tightened properly or completely missing. Such problems are now detected by the latest on-board diagnostics (OBD-II). VOC emissions associated with liquid leaks have also been found. Fuel-hose deterioration, broken or missing hose clamps, and damaged fuel tanks are among the sources of leaks. Indications are that vehicles with leaks, although small in number, can exceed the evaporative emissions of a corresponding nonleaking vehicle by 1 or 2 orders of magnitude (GAO 1997). A half teaspoon of leaked fuel represents more VOCs emitted than exit the tailpipe of a car meeting the current federal standard as it travels over the LA-4 driving schedule (the first two segments of the FTP).

An inherent shortcoming of deterioration studies conducted on large samples of in-service vehicles, whether aimed at exhaust or evaporative emissions, is that they must be restricted to vehicles that have been driven by the public for enough years for malfunctions to develop. Such studies do not account for the effects of future improvements. First, as existing technologies mature, failure rates decrease. Second, gasoline improvements such as reduced volatility and sulfur content decrease emissions (see Chapter 5). Third, new technologies continually emerge. In this latter category is the second generation of on-board diagnostics.

The modern emissions-control system relies on a large number of sensors to provide inputs to the electronic control module. It is important that these sensors operate as intended. First-generation on-board diagnostic capability was introduced with the closed-loop control system in the early 1980s. It checks the function of such key sensors as those measuring coolant temperature, mass airflow, manifold absolute pressure, and throttle position. Malfunction of any of them illuminates a "Service Engine Soon" light on the dashboard, signaling the driver to have the indicated fault corrected by a technician. Second-generation on-board diagnostics, mandated for all model-year 1996 and later LDVs, adds functionality checks on important emissions-control subsystems. For example, the instantaneous acceleration rate of the flywheel is measured to signal a misfiring spark plug. The evaporative emission control system is checked for leaks. Satisfactory function of the oxygen sensor is verified and another check ensures that the EGR valve is working. Comparing signals from oxygen sensors located upstream and downstream of the catalytic converter monitors its effectiveness. EPA expects such advanced early-warning diagnostics to lead to prompt correction of previously undetected malfunctions.

FUNCTIONALITY OF CATALYSTS, OXYGENATED FUELS, AND EXHAUST EMISSIONS

Modern LDVs are equipped with a three-way catalyst that, with proper control of the air-to-fuel ratio via the oxygen sensor, promotes oxidation of most CO and VOCs in the tailpipe to CO_2 and H_2O and reduction of most NO_x to N_2. Moreover, once the catalyst becomes operational, it most readily catalyzes oxidation of the VOC species with the highest reactivity, thus reducing the reactivity of the exhaust stream as well as the total mass of emissions.

Oxygenated fuel makes additional oxygen available to the combustion process and thus, under appropriate circumstances, has the ability to decrease CO emissions. Appropriate circumstances exclude closed-loop operation of the current properly functioning control system because the oxygen sensor adjusts the air-to-fuel ratio to avoid exhaust oxygen. The oxygen sensor cannot discriminate between oxygen molecules in the exhaust that came from air entering the engine and those coming from the fuel itself. On the other hand, during open-loop operation, as during a cold start or a full-throttle acceleration, a system that meters fuel in proportion to airflow without a signal from the exhaust oxygen sensor will supplement the oxygen in the intake air with the additional oxygen in the oxygenated fuel. As this effectively makes the mixture leaner, some reduction in exhaust CO can be expected. (Chapters 6 and 7 discuss effects of the presence of oxygenates in the fuel on the total mass of VOC emissions, as well as the reactivity of those emissions.)

A large fraction of total emissions from the LDV fleet is now known to arise from vehicles without properly functioning emissions control systems. Some of these high emitters might suffer from an improperly functioning catalytic converter. In the limit of a complete failure of the catalyst, the composition of the vehicle tailpipe exhaust approaches the composition of the engine exhaust, and the oxygenated fuel may decrease engine-out VOC and CO emissions somewhat. However, fueling such a high-emitter with an oxygenated fuel will not compensate completely for the loss of conversion efficiency in the catalyst. Also, if the evaporative emissions control system is defective, the higher RVP typical of ethanol-blended RFG might increase evaporative emissions to a greater extent than would a fuel having a lower RVP. The significance of malfunctioning evaporative systems to total VOC emissions from the vehicle has not been studied extensively. The contribution of high emitters is expected to increase in the coming decade (Sawyer et al. 1998).

SUMMARY

Regulatory control of emissions from LDVs has become ever more stringent since the passage of the Clean Air Act. Manufacturers have responded with appropriate control technologies that have become more effective as they mature.

The contribution of ozone precursors to the national emissions inventory by on-road vehicles has been trending downward, despite a substantial increase in vehicle miles traveled.

Evidence indicates that the proportion of driving time spent in transient driving maneuvers that depart significantly from those accounted for by the current FTP is small, but those departures can contribute a disproportionate share of tailpipe emissions. Of particular concern are emissions arising from cold starts and trips with multiple stopovers. Changes to the certification procedure to account for shortcomings of the FTP are forthcoming. However, ongoing monitoring of driving profiles is warranted to ensure realistic integration of tailpipe emissions measurement with the presently uncoupled measurement of evaporative emissions.

The relatively small proportion of high-emitting vehicles present in the fleet can add disproportionately to total fleet emissions. Effectively repairing such vehicles or removing them from the fleet could be the single most effective ozone-precursor reduction strategy in the mobile-source control arsenal. The new generation of on-board diagnostics is expected to decrease the incidence of high emitters. Meanwhile, an ongoing program that diagnoses the specific emissions-component malfunctions in high-emitting vehicles could help to avoid high emitters in the future.

5

Reformulation of Gasoline

FOR THE VAST MAJORITY of light-duty vehicles (cars and smaller trucks), whose engines are spark-ignited, the propulsion fuel is gasoline. As described in Chapter 4, the properties of vehicles and how they are driven influence the quantity of emissions they can generate. The make-up of the fuel that powers a given vehicle can also have a major impact on the emissions, from both a mass and a component speciation point of view. By extension, such a change could exacerbate or mitigate the effects of chronic human exposure to primary and secondary mobile-source air pollution. A brief overview of the toxicology of several oxygenates in fuels is presented in Text Box 5-1.

This chapter reviews recent state- and federal-regulatory efforts to protect human health and the environment by means of the modification, or reformulation, of motor gasoline—emphasizing the requirements for the addition of oxygen and how that oxygen is provided. A discussion then follows on the properties and laboratory-measured performance of these reformulated gasolines (RFGs) with respect to the amount of ozone precursor emissions (volatile organic compounds (VOCs), oxides of nitrogen (NO_x), and carbon monoxide (CO)) generated by the vehicles that use them. This discussion is intended to serve as preface for the review of actual in-use case studies and real-world observations of air-quality effects presented in Chapter 6.

TEXT BOX 5-1 Toxicological Considerations of Oxygenates in Fuels

Although this report focuses on the effects of motor-vehicle fuel composition on formation of tropospheric ozone, earlier reports dealt in considerable detail with the toxicological and health effects related to fuel composition. Two reports that focused specifically on the effects of oxygenates in fuels are *Toxicological and Performance Aspects of Oxygenated Motor Fuels* (NRC 1996) and *Interagency Assessment of Oxygenated Fuels* (NSTC 1997).

The NRC report reviewed a draft of the interagency assessment, and recommended a number of refinements and improvements in the assessment of potential human health risks associated with prolonged exposure to gasoline containing MBTE and in the assessment of the comparative risks associated with oxygenated and nonoxygenated fuels. The NRC report concluded that "until these recommendations are acted upon, no definitive statement can be made regarding these health-risk issues. Based on the available analysis, however, it does not appear that MTBE exposure resulting from the use of oxygenated fuels is likely to pose a substantial human health risk. It appears that MTBE-containing fuels do not pose health risks substantially different from those associated with nonoxygenated fuels, but this conclusion is less well established and should become the centerpiece for the government's comprehensive assessment."

The interagency assessment report concluded that "it is not likely that the health effects associated with ingestion of moderate to large quantities of ethanol would occur from inhalation of ethanol at ambient levels to which most people may be exposed from use of ethanol as a fuel oxygenate. Potential health effects from exposure to other oxygenates are not known and require investigation if their use in fuels is to become widespread."

In a related issue, in 1998, U.S. Senator Dianne Feinstein (California) requested an investigation of possible contamination of the nation's groundwater by MTBE and sought help from EPA in dealing with potentially serious MTBE issues confronting California, namely, water contamination in the state. Moreover, Senator Barbara Boxer had requested that EPA phase out MTBE because of mounting evidence of MBTE contamination of California's drinking water. EPA announced in November 1998 that it will undertake a pilot site-remediation demonstration project in California. On March 25, 1999, California Governor Gray Davis issued Executive Order D-5-99, which requires the phase out of MTBE from California gasoline by no later than December 31, 2002.

BASIC PROPERTIES

Irrespective of any regulation of their content, the composition and properties of motor-vehicle fuels are routinely tailored to meet the requirements of the existing and the emerging fleet of automobiles and trucks. Cost and general availability are obviously major considerations; in the past, a fuel's specifications were established by vehicle manufacturers, together with the fuel's producers. Increased concern about air pollution and health effects from the use of motor-vehicle fuels have brought the federal and some state governments, through their environmental regulatory agencies, into an increasingly prominent role in determining fuel composition. Within this regulatory context, fuel composition has typically been defined by specifications set as a range of properties, each having a maximum or minimum or both stipulated.

Volatility and Distillation Curve

Fuel volatility and distillation are related to the composition of vapors in the gasoline tank and in the fuel delivery system. They are critical to the proper operation of the engine. For example, a sufficiently high "front end volatility" is required for cold starting a vehicle and is generally higher in the winter than in the summer. Fuel volatility is often expressed in terms of the Reid vapor pressure (RVP), which is defined as the vapor pressure (or gauge pressure) of a liquid at 100°F, as measured in a standardized apparatus, or "bomb" (in pounds per square inch (psi)). A distillation curve can be characterized by the temperatures (usually in °F) at which 10%, 50%, and 90%, respectively, of the fuel is distilled (or evaporated). Those temperatures are represented by T_{10}, T_{50}, and T_{90}.

Octane Number

Octane number is a measure of the tendency of a fuel to detonate during combustion in a standardized variable-compression-ratio "knock"-test engine in which the compression ratio[1] is increased until knock is de-

---

[1]"Compression ratio" is the ratio of the volume of a cylinder with a piston at bottom dead center to the volume of that cylinder with the piston at bottom dead center.

tected. The test results for a fuel are scaled to an octane number of zero for n-heptane and 100 for iso-octane (2,2,4-trimethylpentane). The sets of measurement conditions generally applied for determining each octane-rating component are summarized in Table 5-1.

With the phasing out of tetraethyl lead from motor gasoline, changes in composition were necessary to maintain the octane number of the unleaded gasoline so that the current and future fleets of passenger cars could operate properly. This was accomplished by increasing the content of high-octane hydrocarbons such as alkylated aromatics, olefins, and branched paraffins. Oxygenated compounds (e.g., alcohols and ethers) are also high-octane blending components, and their use as octane enhancers began as early as the late 1960s.

Oxygenates in Fuels

The major components of gasoline are hydrocarbons, whose elemental make-up includes only carbon and hydrogen. For a variety of reasons, including a desire to minimize motor-vehicle pollutant emissions, a small amount of chemically-combined oxygen is sometimes incorporated into the fuel by adding an oxygenated organic compound to the blend. The two oxygenated compounds most commonly used as additives in gasoline today are MTBE ($CH_3OC(CH_3)_3$) and ethanol (C_2H_5OH).

The amount of oxygen in a fuel is usually expressed in terms of the percent of oxygen in the fuel by weight (i.e., wt % oxygen) or the percent by volume of the oxygenated additive (i.e., vol % additive). Table 5-2 presents the values for vol % of ethanol and MTBE that correspond to a range of wt % oxygen contents that are typical of RFG blends. Note that because ethanol contains more oxygen on a per-gram basis than

TABLE 5-1   Test Parameters for Octane Measurement

	Research Octane Number (R)	Motor Octane Number (M)
Engine speed, rpm	600	900
Air temperature, °F	60-125	100
Usefulness	Provides relative numbers for low-speed, mild-knock conditions	Provides relative numbers for high-speed, high-knock conditions

NOTE: The antiknock index, 0.5 (R + M), is commonly used.

TABLE 5-2   Amounts of Ethanol and MTBE Needed to Produce a Given Oxygen
Content in RFG

Wt % Oxygen	Vol % Ethanol	Vol % MTBE
1	2.85	5.6
1.5	4.3	8.3
2	5.7	11.2
2.5	7.1	13.9
3.0	8.6	16.7
3.5	10.1	18.9

does MTBE, about 50% less ethanol (by volume) is required to produce
a given % wt of oxygen in a fuel than in the case of MTBE. As discussed
later in this chapter, the federal RFG program mandates a minimum 2 wt
% oxygen in all RFG blends. In Table 5-2, it is shown that meeting such
a requirement takes a little less than 6 vol % ethanol and a little more
than 11 vol % MTBE. It turns out however, that when ethanol is present
in fuel at concentrations of a few vol % to about 10 vol %, it tends to
significantly enhance the fuel's RVP.[2] As a result, there is a general
tendency for ethanol-containing blends to contain more oxygen (on a wt
% basis) than MTBE-containing blends.

Because the octane numbers for both ethanol and MTBE are relatively high, they are attractive additives for use in lead-free gasoline.
Except where mandated by law, however, oxygenate producers compete
with conventional refining processes for producing high-octane hydrocarbons that can be added to gasoline. These conventional processes include
the following:

•   Catalytic cracking to increase the amount of components with
boiling points in the range of that of gasoline and to produce high-octane
olefins and aromatics.

•   Catalytic reforming to convert naphthenes and some paraffins
to high-octane aromatics.

---

[2]Studies indicate that fuel RVP increases as ethanol is initially added. The
greatest RVP increase occurs with an ethanol content of about 5 vol % and is
about 1 psi. For ethanol concentrations greater than 5 vol %, the RVP slowly
decreases.

- Isomerization and alkylation to produce branched paraffins.

In general, those processes can be more economical than those that produce oxygenates; and thus, oxygenates were not initially the additive of choice for enhancing octane number in fuels, as discussed later in this chapter. However, in addition to enhancing octane number, oxygenates in gasoline can provide air-quality benefits. For example, as discussed in Chapter 4, use of oxygenates can lower emissions of CO during open-loop operation (such as warm up) in modern vehicles (i.e., those with closed-loop feedback control) and in vehicles that do not have closed-loop controls. There is also some indication that oxygenates can lower the mass and reactivity of VOC exhaust emissions in some cases (see Chapters 6 and 7). The presence of oxygenates in reformulated gasoline has been mandated by law and regulation, and this provides the incentive for using oxygenates to boost octane number instead of using components produced by conventional processes.

All things being equal, the choice of which specific oxygenate to use would be dictated by economic factors; that is, which oxygenate can produce the desired gasoline characteristics (e.g., high-octane number) at the least cost. The principal production method for ethanol used in gasoline is fermentation of carbohydrates from grain (mostly corn):

carbohydrates (e.g., sugars) + yeast $\rightarrow$ ethanol (C_2H_5OH) + residue.

Ethanol is also produced in petrochemical facilities through ethane-ethene synthesis:

$$C_2H_6 \rightarrow C_2H_4 + H_2$$
$$C_2H_4 + H_2O \rightarrow C_2H_5OH.$$

MTBE, on the other hand, is produced in a two-step process, with petrochemical synthesis employed to manufacture methanol from natural gas:

$$CH_4 + H_2O \rightarrow CO + 3H_2$$
$$CO + 2H_2 \rightarrow CH_3OH.$$

2-Methylpropene is manufactured from 2-methylpropane:

$$CH_3CH(CH_3)_2 \rightarrow CH_2 = C(CH_3)_2 + H_2.$$

MTBE is then produced by reacting methanol with 2-methylpropene:

$$CH_3OH + CH_2 = C(CH_3)_2 \rightarrow CH_3OC(CH_3)_3.$$

This multistep process makes use of readily available inexpensive feed-stock and enables MTBE to be produced at a cost that is generally less than that of producing ethanol by grain fermentation. However, in the United States, tax subsidies have made ethanol production via fermentation competitive with MTBE production. Because the committee was not asked to address this aspect of the RFG issue, the economic implications of using MTBE versus ethanol as an oxygenated additive are not discussed in this report. A discussion of the potential air-quality benefits of the two oxygenates is presented in Chapter 7.

Sulfur in Gasoline

Sulfur (combined chemically in the organic components of the fuel) is a trace impurity of gasoline. Reductions in gasoline sulfur content can substantially improve catalytic-converter performance (AQIRP 1992), as well as lower sulfur dioxide (SO_2) emissions. Sulfur's effect in impairing the function of a catalytic converter by poisoning the catalyst is believed to be reversible. Removal of sulfur to a low weight-percent of gasoline (i.e., ≤ 100 parts per million (ppm) by weight) can be accomplished by hydro-desulfurization of catalytic, thermal, and virgin naphtha.

FEDERAL AND CALIFORNIA REGULATION OF GASOLINE PROPERTIES

History of Federal Actions Before 1994

The first federally mandated gasoline reformulation in recent history was the staged removal of the octane-enhancing additive tetraethyl lead from all motor gasolines. In general, the function of the oxidizing exhaust catalyst of a vehicle is impaired when the vehicle is operated with leaded gasolines. In anticipation of the introduction of catalysts to the light-duty motor-vehicle fleet in 1975, the U.S. Environmental Protection Agency (EPA) began phasing out leaded gasoline in the early 1970s (EPA

1973). A subsequent EPA rule restricted the lead content of any gasoline to a maximum 0.1 grams per gallon (g/gal) as of January 1, 1996, to achieve reductions in the inhalation exposure of humans (especially young children) residing in urban areas to airborne lead. Up to 1995, trace amounts of lead (up to 0.05 g/gal) could still be included in gasolines, but thereafter gasolines in the United States were mandated to be essentially lead-free.

Because lead had been in gasoline for many years to enhance combustion performance (by increasing its octane rating or antiknock index), a comparably effective substitute additive was desired. Initially, lower paraffins, such as butane, offered the combination of octane enhancement and cost-effectiveness that refiners sought because they boosted the rating sufficiently at relatively low concentrations. However, butane in particular evaporated readily, having an RVP of about 58 psi and also volatilized other reactive hydrocarbons in the gasoline. The result was an industry-average gasoline with an RVP as much as 2 to 3 psi higher during the ozone season than that of the EPA certification test gasoline.

Through about 1987, discrepant volatility was not an issue because excursions of the 1-hr ambient ozone concentration standard of 0.12 ppm in most locations had been in steady decline. However, the summer of 1988 witnessed some of the worst ozone excursions on record (see Chapters 4 and 6). These excursions were widespread and often of long duration because of unusually protracted hot and sunny conditions and air stagnation over much of the nation. The ozone excursions led to speculation that evaporation of the then-common high-volatility summer gasoline, in use and in bulk storage, was a major contributor to the mass of VOC emissions giving rise to these ozone episodes. A seminal compendium of peer-reviewed research results, at that time, identified reduction of gasoline volatility as the most effective means then available to reduce anthropogenic VOC emissions attributable to mobile-source activity (NAPAP 1991).

The air-regulatory structure created under the National Environmental Protection Act (Public Law 91-190) of 1969 and the Clean Air Act (CAA) Amendments of 1970 had sought to substantially reduce transportation's contribution to the ozone problem through an almost exclusive programmatic focus on motor-vehicle manufacturers (Chapter 4). The core of this structure was a set of increasingly stringent per-vehicle emissions standards (shown in Tables 4-1 and 4-2) called the Federal Motor Vehicle Control Program. Beginning in 1989, the structure expanded to encompass the fuels industry, especially petroleum producers, in the quest for greater control of emissions from gasoline-powered vehicles. Following initiatives taken by individual states, such as Colorado, EPA promulgated a rule that set upper RVP limits for gasoline sold during the ozone season throughout the nation (EPA 1989). The limits were determined, in part by meteorology, but largely by average summer temperatures. These limits were subsequently redefined and made more stringent for 1992 and later years (EPA 1990). This initial foray by the federal government into using fuel properties to aid in ozone mitigation efforts was then substantially expanded by the passage of the Clean Air Act Amendments of 1990, which mandated the federal RFG program. The key aspects of this program are discussed later in this chapter.

Corresponding California Actions

Various regions of California exceed the air-quality standards for ozone several times per year, and the Los Angeles area is generally recognized as having the most severe ozone pollution problems in the nation. Perhaps, for this reason, California has often led the nation in the promulgation of new and creative approaches to ozone-pollution mitigation, and regulation of gasoline is no exception. Requirements for fuel modifications in California have existed since 1971 when RVP limits were mandated. Through the 1970s, requirements were also promulgated for quantities of lead, sulfur, and manganese-phosphorous in gasoline and sulfur in diesel fuels.

The California Clean Air Act of 1988 imposed additional requirements on mobile sources to (1) achieve maximum emissions reductions of VOCs and NO_x by the earliest practicable date; (2) achieve feasible reductions in particulate mass (PM), CO, and toxic-air contaminants; and (3) adopt the most effective control measures on all classes of motor vehicles and their fuels. In response to this, the California Air Resources Board (CARB) adopted the California RFG regulations to require cleaner-burning gasoline. This program is a critical component of California's State Implementation Plan (SIP) to reduce air pollution, and will also meet the requirements of the federal RFG program some 3 to 4 years earlier than that mandated in the CAA Amendments of 1990. Motor-vehicle-exhaust emissions standards were further specified under California's Low Emission Vehicles and Clean Fuels Program.

The Auto/Oil Study

A key principle first manifested in the concept of an RFG program is the concept that a vehicle and its fuel are an integrated system for which emissions controls should be fashioned to derive the optimum benefit from each of the system's components. In acknowledgment of this principle, the auto and oil industries initiated the Auto/Oil Air Quality Improvement Research Program (AQIRP) in 1989. The purpose of AQIRP was to develop data on potential improvements in vehicular emissions and air quality that could be realized through the use of RFG, various alternative fuels, and the development of automotive technology (Burns et al. 1992).[3]

AQIRP sought to identify those fuels and formulations that could be most effective in reducing ozone precursors without compromising driveability or substantially increasing the cost (per gasoline- or diesel-equivalent range) of driving. The program was motivated in part by the perception that the crafting of gasoline should be completely rethought, such that the entire range of its potentially health-harmful constituents, including sulfur, aromatics, and reactive olefins, should be subject to limits. The AQIRP findings have served as the cornerstone for the design of both the federal and California RFG programs, and are discussed in depth in Chapter 6.

WHAT IS REFORMULATED GASOLINE?

There are currently in the United States two RFG programs: a federal program mandated in Section 211(k) of the CAA and a California program. The California program precedes the federal program by about 3 to 4 years. Both the federal and California programs are to be implemented in two phases. (To avoid confusion, Arabic numerals are used in this report to identify Phases 1 and 2 of the California program, and Roman numerals are used to identify Phases I and II of the federal program.) The general characteristics of the two programs are outlined

---

[3] Three U.S. automobile companies (Ford, General Motors, and Chrysler) and 14 petroleum companies (Amoco, ARCO, Ashland, BP, Chevron, Conoco, Exxon, Marathon, Mobil, Phillips, Shell, Sunoco, Texaco, and Unocal) planned and carried out AQIRP.

in Table 5-3 Parts 1 and 2. (The tables are not intended to provide a comprehensive presentation of the programs' requirements.)

The federal and California RFG programs are specifically aimed at mitigation of the ozone-pollution problem through the reduction of light-duty-vehicle (LDV) emissions of VOCs, CO, and NO_x. These programs should not be confused with oxygenated fuels programs, such as the Federal Oxygenated Fuels Program (see Table 5-4), which seeks to lower motor-vehicle emissions of CO to avoid nonattainment of the National Ambient Air Quality Standard (NAAQS) for CO. Because CO pollution is typically most severe in the winter months, the oxygenated fuels program generally seeks to regulate fuel composition during those months. By contrast, the RFG programs tend to prescribe content and volatility of gasoline sold during the summer ozone season.

Federal RFG Program

In general terms, the federal concept of RFG, as of January 1, 1998, is gasoline blended such that, on average, the exhaust and evaporative emissions of VOCs and air toxics (chiefly benzene, 1,3-butadiene, polycyclic organic matter (POM), formaldehyde, and acetaldehyde) resulting from RFG use in motor vehicles are significantly and consistently lower than such emissions resulting from use of conventional gasolines. In a legal context, a gasoline is reformulated if the EPA administrator has certified that it meets all specifications of the CAA. Section 211 of the CAA codifies the redefinition of gasoline to be sold in areas failing to achieve ambient air-quality standards for air pollutants linked to emissions of CO, nonmethane hydrocarbons (NMHCs), and NO_x. As described in Chapter 2, all three are precursors for tropospheric ozone formation. (CO and nitrogen dioxide (NO_2) are also subject to ambient-concentration standards because of their direct impact on human health.)

As indicated in Table 5-3 Part 1, nine metropolitan areas are specified for application of the federal RFG program. Before passage of the CAA Amendments of 1990 that codified these requirements, EPA had already concluded that those areas would require an arsenal of new weapons to combat their ozone problems, and that changes in the composition of motor fuels would play a key role. Subsection 211(k) (10)(D) officially defined those areas as the "covered areas" for use of

TABLE 5-3 Part 1: California and Federal Reformulated Gasoline Programs[a]

California RFG Program, Phase 1 (1992-1996)	Federal RFG Program, Phase I (1995-1999)
• Effective January 1, 1992. • Set gasoline RVP limit at 7.8 psi. • Required detergent additives and no lead in gasoline. • *No explicit oxygen requirement for summer gas.*	• Mandated in 42 U.S.C. 7545 as a result of language in Section 211(k) of the CAA Amendments of 1990. • Effective beginning 1/1/95 in the 9 metro ozone-nonattainment areas with population of 250,000 or greater classified as "extreme" or "severe" as of 11/15/90: • Los Angeles (South Coast Air Basin) • San Diego • Baltimore/Washington • Hartford-New Haven-Waterbury, CT • New York/New Jersey/SW Connecticut • Philadelphia/Wilmington/Trenton • Chicago/NW Indiana • Milwaukee/Racine, WI • Houston/Galveston/Brazoria, TX (Sacramento, CA was later added)
California RFG Program, Phase 2 (1996-) • Effective with beginning of 1996 ozone season. • Set flat limits for the following properties: • RVP: 7.0 psi (gauge) • Sulfur: 40 ppm (vol) • Oxygen: 0-2.7% (wt) • Olefins: 6.0% (vol) • Aromatics: 25% (vol) • Benzene: 1.0% (vol) • Temperature at which 50% of fuel is distilled/vaporized (T_{50}): 210°F. • Temperature at which 90% of fuel is distilled/vaporized (T_{90}): 300°F. • Meets federal Phase II RFG specification and performance requirements (see Table 5-3 Part 2) *except* that oxygenate content requirement may be waived if a refiner demonstrates, through emissions test results for 20 vehicles in four technology classes, that a fuel's exhaust-emissions performance targets can be achieved without it.	• Specified content criteria for gasolines to be sold in these areas primarily during the *summer* ozone season: oxygen minimum of 2.0% by wt; benzene maximum of 1.0% by vol; aromatics maximum of 25.0% by vol; must contain detergent additive; must exclude heavy metals. • Per-gallon performance requirements: 15.0% reduction in toxics; at least 15.6% northern states; 35.1% southern states; reduction in VOC relative to specified baseline gasoline, as computed by Complex Model (Simple Model valid until 1/1/98).

113

• Properties *may* be measured according to an average limits provision, as long as the flat limits are met on average over a specified period of time.
• RFG performance relative to that of a specified base fuel *for exhaust emissions only* is calculated with the Predictive Model, which California developed using approximately the same data base that EPA used in developing the Complex Model.

• Average performance requirements (across all RFGs from a refiner): at least 16.5% reduction in toxics; at least 17.1% northern states, 36.6% southern states; reduction in VOCs as computed by Complex Model (Simple Model valid until 1/1/98).
• RVP limits based on 40 CFR 80.28 standards, which cover all gasolines sold. Other areas may opt in to program irrespective of ozone attainment status and may opt out if alternative means of attaining (and maintaining) ambient ozone standards are demonstrated.

aUnless otherwise stated, standards for the first phase of both programs carry forward to the second phase.

114

TABLE 5-3 Part 2: Future Reformulated Gasoline Program

Federal RFG Program, Phase II
(2000-)

• Effective January 1, 2000.
• Revises per-gallon performance criteria for gasolines to be sold in covered and opt-in areas during the ozone season: at least 20% reduction in toxics; at least 25.9% (northern states) and 27.5% (southern states) reduction in VOCs; and at least 5.5% reduction in NO_x (which was not previously controlled) for VOC-controlled areas; relative to specified baseline gasoline, as computed by the Complex Model.[a] (See Table 5-6.)
• Similarly, if a refiner opts to meet performance criteria on a pooled average (rather than per-gallon) basis as described in 40 CFR 80.67, targets are at least 21.5% reduction in toxics; at least 27.4% (northern states) and 29.0% (southern states) reduction in VOCs (but 23.4% and 25.0%, respectively, for any individual gallon sampled); and at least 6.8% reduction in NO_x for VOC-controlled areas; all relative to specified average baseline gasoline, as computed by the Complex Model.
• For areas not designated VOC-controlled,[a] the pooled average NO_x reduction standard for RFGs is 1.5%.
• Per-gallon oxygen minimum requirement relaxes to 1.5% by wt as long as an average oxygen content across all RFGs produced by a refiner for a given area is 2.1% or higher.
• Per-gallon benzene maximum requirement relaxes to 1.3% by wt, as long as an average benzene content across all RFGs produced by a refiner for a given area is 0.95% or lower.

[a]Opt-in areas are listed in Table 5-5.

TABLE 5-4 Federal Oxygenated Fuels (Oxyfuels) Program (1992-)

- Mandated in 42 U.S.C. 7545 as a result of language in Section 211(m) of the CAA Amendments of 1990.
- Justified based on the apparent success of winter gasoline oxygenation programs established from 1988 onward by states such as Colorado.
- Effective no later than 11/1/92 for any area with a CO design value (nonattainment ambient concentration level) of 9.5 ppm and above as of 11/15/90 (on 11/1/92 there were a total of 23 qualifying areas, excluding the 8 in California that were separately controlled by CARB regulation from November 1992).[a]
- Requires gasoline sold in such areas during the winter (high-CO) season to contain not less than 2.7% oxygen by wt (over a minimum period of 4 months) but controls no other parameter.[b]
- Provides, pursuant to CAA Section 211(m)(2), that the EPA Administrator may waive these requirements for any area in which they would interfere with the attainment of a national, state, or local ambient air-quality standard for any pollutant other than CO.
- Otherwise, there is no "sunset" provision for the program, except for any qualifying area that attains the CO standard and demonstrates that it can maintain it without the use of oxygenates.

[a]California began using oxygenates in November 1992 to comply with federal requirements for the control of CO, and implemented a modified form of the Federal Oxygenated Fuels Program to reduce ambient CO in 1992 in approximately 40 nonattainment areas (Kirchstetter et al. 1996). The periods in which oxygenates were required depended on location, but all were in the range of October 1 through February 29.

[b]California's adopted requirement is lower at 1.8% to 2.2% due to concern about NO_x formation.

summer RFG. In addition, the subsection allowed for participation of any other nonattainment areas wishing to opt in to the gasoline content regulations. Table 5-5 lists these so-called RFG voluntary opt-in areas as of May 1, 1998.

The intent of Congress in formulating the federal RFG program was to ensure the participation of oxygen and oxygenate constituents in the modified-gasoline composition. The addition of oxygen at a minimum of 2.7% content by weight to winter-blended gasoline as a means to control CO emissions was required on a national basis for CO nonattainment areas outside California in CAA Section 211(m)(2). Further, as part of the federal RFG program, the EPA Administrator was instructed by CAA Section 211(k)(2)(B) to set content requirements for oxygen, benzene, and aromatics in any gasolines. For oxygen, the section requires a minimum of 2.0% by weight; for benzene, a maximum of 1.0% by volume; and, for aromatics, a maximum of 25% by volume. Such gasolines were also prohibited from containing heavy metals and from excluding detergent additives. Subsection 211(k)(7) also provides for a determination of credits for refining gasolines certified to a greater stringency (i.e., less benzene or aromatics, more oxygen) than that stipulated by the stated limits.

In developing regulations to implement the requirements of the CAA Amendments of 1990, specific per-gallon emissions-reduction targets for RFGs were set for each of two VOC-control regions: those generally in the northern tier of states, with summer baseline gasolines in the 9.0-psi RVP range, and those generally in the southern tier, where summer gasoline has RVPs of 8.0 psi and lower. Such a distinction reflects the

TABLE 5-5   Federal RFG Opt-in Areas As of July 1998

Connecticut (entire state)	Massachusetts (entire state)
Delaware (entire state)	Boston-Lawrence-Worcester portion,
Washington, DC (including MD and	MA and NH
VA suburbs)	New Jersey (entire state)
Louisville, KY	Duchess and Essex Counties, NY
Auburn and Portland, ME	Rhode Island (entire state)
Lewiston, Knox, and Lincoln	Dallas-Ft. Worth, TX
Counties, ME	Richmond, VA
Kent and Queen Anne's Counties,	Norfolk, Virginia Beach, and
MD	Newport News, VA

historical industrial practice where southern gasoline had lower RVP than northern gasoline to compensate for higher ambient temperatures. (The covered areas outside California that fall into the southern tier include only the Houston-Galveston-Brazoria, Texas area, and the recent opt-in areas of Dallas-Ft. Worth, Texas; and the Richmond and Tidewater metro areas of Virginia.) A target of 15% reduction for toxics and at least 15.6% for VOCs was defined for northern-tier ozone control regions; the southern-tier VOC-reduction target was 35.1%. As described below, these targets are to be computed using the so-called "Complex Model." Following the mandatory introduction of Phase II RFG after December 31, 1999, the targets rise to 20% minimum reduction for toxics and 25.9% reduction of VOCs (or at least 23.4% on a year-round averaged basis for certified RFGs) for northern-tier RFGs; and a 27.5% reduction of VOCs (at least 25.0% on a year-round averaged basis) for southern-tier RFGs. Phase II RFG is also required to reduce exhaust emissions of NO_x by 5.5% in both VOC-control regions, relative to a defined baseline gasoline.

California RFG Program

In 1991, shortly after passage of the CAA Amendments of 1990 and the establishment of the federal RFG program, CARB adopted the California RFG regulations to require cleaner-burning gasoline. This program is now a critical component of California's SIP (State Implementation Plan) to reduce air pollution and meet the requirements of the CAA. Phase 1 requirements, effective January 1, 1992, required: (1) an RVP limit of 7.8 psi; (2) deposit-control additives to prevent and reduce deposits; and (3) elimination of leaded gasoline from on-road motor vehicles.

Refiners were required to begin making California Phase 2 RFG in March 1996 (Cal EPA 1996) and the gasoline was introduced statewide in the summer of 1996. As indicated in Table 5-3 Part 1, California Phase 2 RFG was required to limit specific fuel properties: RVP to reduce evaporative VOCs; sulfur content to avoid catalyst poisoning and thereby reduce VOCs, NO_x, CO, and toxic emissions; aromatics content to reduce the atmospheric loading of mono- and polycyclic hydrocarbons linked to ozone formation; olefins to reduce VOC emissions reactivity; and benzene content to decrease emissions of this regulated toxic substance.

Phase 2 also has an oxygen requirement of 1.8% to 2.2% by weight during the winter or CO-control season.[4] (Recall that there is also a minimum winter oxygenate content standard of 1.8% by weight for those areas that have to comply with federal RFG requirements for CO (see Table 5-4)).

Introduction of the federal Phase II RFG program will begin in the year 2000. Because the California Phase 2 RFG program was initiated in 1996 and satisfies the requirements of the federal Phase II program, it offers an opportunity to determine the effect of Phase II RFG before its more general introduction. This aspect of the RFG program is addressed in Chapter 6.

California allows four methods for determining the fuel properties for RFG. First, there is a flat-limit provision (listed in Table 5-2 Part 1) whereby all of the fuels from a given refiner at all times must meet the standard. Second, there is an average-limit provision, whereby the refiner can average gasoline properties over time, but with caps (upper limits) on properties that cannot be exceeded in blending. Third, refiners can produce market-equivalent gasoline formulations evaluated with the California Predictive Model (Cleary 1998). Fourth, refiners can produce market-equivalent formulations evaluated by emissions testing. The emissions-testing option for alternative property limits requires testing 20 vehicles in four technology classes with different vehicles and considering only exhaust emissions. (For reasons of practicality, refiners tend to rely on the Predictive Model instead of emissions testing.) Cap limits and constant RVP were imposed on the property limits. In principle and in contrast to the federal Phase II RFG requirements, a fuel can be certified for Phase 2 of the California RFG program through emissions testing without having any oxygenates.

FEDERAL REQUIREMENTS FOR RFG UNDER THE COMPLEX MODEL

In a final rule of February 16, 1994 (EPA 1994), EPA articulated requirements for RFG blends eligible for sale during the ozone season in the

---

[4]An oxygen content of 2% by weight is equivalent to an MTBE concentration of 11.2% by volume, or an ethanol concentration of 5.7% by volume.

covered and opt-in areas. These were based upon Simple and Complex Models to determine the allowable maxima and minima of regulated constituents. Either model could be used for determining formulation of gasoline for federal RFG through December 31, 1997. After that date, only the Complex Model could be used for Phase I and then for Phase II, which begins in the year 2000. Because the Simple Model is no longer being used, only the Complex Model will be discussed here.

The Complex Model provides values of emissions due to modification of fuel properties with respect to RVP, oxygen by weight-percent, benzene content, sulfur, olefins, and distillation fraction (at 200°F and 300°F). The values are generated by a set of regression relationships derived by EPA from tests on differing fuel formulations performed in the 1980s and early 1990s on gasoline-fueled cars and trucks of numerous model years (and emissions-control technologies). Many hundreds of tests were involved. The regulation defines both summer and winter baseline fuel properties with respect to eight content variables, and from these derives baseline exhaust emissions of VOCs, NO_x, toxics, and polycyclic organic matter, and evaporative emissions of VOCs and benzene as the basis of comparison for both federal Phase I and Phase II RFG. There is further allowance for the difference between normal and high emitters, such that emissions-reduction credit for each season and RFG phase must be weighted between them. The complete set of performance calculation equations for federal Phase II RFG summer emissions in the Complex Model is provided in Appendix C. The appendix shows the various (flat and variable) fuel-content standards and emissions standards of performance to be computed using the Complex Model that Phase II RFG must meet.

The Complex Model is complex because it separates the consideration of fuel properties and the calculation of a fuel's emissions performance into many different classification bins that vary by geographical region, season, RFG phase, emitter category, and range of parameters of fuel constituents. All emissions comparisons rest on EPA's definitions of baseline fuel properties and emissions (Table 5-6).

THE CALIFORNIA PREDICTIVE MODEL

The California Predictive Model is based only upon exhaust-emissions measurements. Because evaporative emissions represent a substantial fraction of the total VOC emissions from LDVs, their omission represents

TABLE 5-6 Base and Example Ozone Season Phase II RFG Requirements

Fuel Parameter	Baseline Value	Qualifying Phase II RFG Value
Benzene (vol %)	1.53	≤1.00
Oxygen Content (wt %)	0.0	≥2.0
RVP (psi)	8.7	6.8 (expected average)
Aromatics content (vol %)	32.0	—[a]
Sulfur (ppm)	339	—[a]
Olefins (vol %)	9.2	—[a]
200°F distillation fraction	0.41	—[a]
300°F distillation fraction	0.83	—[a]

Emission category	Baseline value (mg/mi)	Required % reduction from baseline computed from Complex Model for (a) per gallon or (b) pooled average over all of any refiner's RFG output
Exhaust VOCs	907.0	((a) ≥27.5, (b) ≥29.0 (southern)) sum of exhaust VOCs + nonexhaust[a] VOCs
Nonexhaust VOCs	559.31 (for southern states) 492.07 (for northern states)	((a) ≥25.9, (b) ≥27.4 (northern)) sum of exhaust VOCs + nonexhaust[a] VOCs
NO_x	1,340.0	NO_x minimum reduction requirements removed effective 1/1/98
Exhaust benzene	53.54	—[c]
Nonexhaust benzene	6.24 (southern) 5.50 (northern)	—[c]
Acetaldehyde	4.44	—[c]
Formaldehyde	9.70	—[c]
1,3-Butadiene	9.38	—[c]

Polycyclic organic matter	3.04	—[c]

Additional Fuel-Content Requirements

Oxygen content (wt %)	(a) $\geq$2.0; (b) $\geq$2.1
Per-gallon minimum O_2 under option b (wt %)	1.5
Year-round maximum O_2 (wt %)	2.7 (MTBE)
	3.5 (ethanol)
Benzene content (max. vol %)	(a) 1.00; (b) 0.95
Per-gallon maximum benzene under option b (vol %)	1.5

[a]Any combination of aromatics content, sulfur, olefins, 200°F distillation fraction, or 300°F distillation fraction, that collectively results in the target fuel meeting the performance levels for the pollutants shown in the table.

[b]If option b is selected, per-gallon percent reduction requirements of 25.0 (southern states) and 23.4 (northern states) still apply.

[c]Collective percent reduction requirement for benzene in exhaust and nonexhaust, acetaldehyde, formaldehyde, 1,3-butadiene, and polycyclic organic matter is $\geq$20.0 per gallon or $\geq$21.5 as a pooled average.

a serious limitation of this model. The model is derived from data collected from 20 different test programs that investigated the relationship between fuel properties and exhaust emissions. In the course of these studies, over 1,000 vehicles were tested using 200 different fuels. In spite of the rather large numbers, many fuels were evaluated on a rather small set of vehicles. Only two vehicle types were modeled: 1980-1985 model years (i.e., the Tech 3 class), and post-1985 model years, (the Tech 4 class). Caps limited the range of fuel-property values, and RVP was held constant and not treated as a variable in the regression formula. Perhaps holding RVP constant was the basis for the neglect of evaporative emissions; however, neglecting those emissions biases the overall emissions estimates for the vehicle fleet. The resulting model consists of a series of regression equations that describe the exhaust emissions of NO_x, VOCs, and potency-weighted toxics as a function of various properties of the fuel blend. For example, the NO_x and VOC emissions for the Tech 4 class (in units of percent reduction from a California Phase II reference fuel) are given by

$$\text{Tech 4 } (NO_x) = A + B(T_{50}) + C(T_{90}) + D(Aromatics) + E(Olefins) + F(O_2) + G(Sulfur) + H(Aromatics \times O_2) + I(O_2)^2,$$

$$\text{Tech 4 (VOC)} = A + B(T_{50}) + C(T_{90}) + D(Aromatics) + E(Olefins) + F(O_2) + G(Sulfur) + H(T_{50} \times T_{90}) + I(T_{90} \times T_{90}) + J(T_{90} \times O_2) + K(T_{90} \times Aromatics) + L(Aromatics)^2,$$

where the coefficients are given in Table 5-7.

There are different formulas for Tech 3 vehicles; these are not included here for brevity. For a fuel to qualify for the program, the model-predicted emissions for the proposed blends must be less than the California default limits. Refiners use the model to validate their blends and to adjust limits of fuel properties to fit refinery operations.

PERFORMANCE AND RELIABILITY OF COMPLEX AND PREDICTIVE MODELS

The Complex Model and Predictive Model were designed to predict reductions in the mobile-source emissions of NO_x, VOC, and toxics as a result of the use of RFG. The models are used to certify a candidate fuel for the federal RFG program (Complex Model) or California RFG program (Predictive Model). Because of their limitations as discussed in this

TABLE 5-7 Coefficients for the California Predictive Model for Tech 4 Vehicles

Coefficient	Tech 4(NO_x)	Tech 4(VOC)
A	6.82×10^{-1}	-1.16
B	1.95×10^{-3}	7.64×10^{-2}
C	-8.20×10^{-3}	3.89×10^{-2}
D	4.14×10^{-3}	1.37×10^{-1}
E	2.59×10^{-2}	-6.87×10^{-3}
F	-8.99×10^{-3}	-1.04×10^{-2}
G	5.01×10^{-2}	1.17×10^{-1}
H	-5.79×10^{-3}	2.58×10^{-2}
I	1.35×10^{-2}	1.82×10^{-2}
J		1.51×10^{-2}
K		1.21×10^{-2}
L		-1.20×10^{-2}

report, these models are not used routinely to generate input data for regulatory air-quality models to assess the ozone reductions. In Chapter 7 of this report, the Complex and Predictive Models were used to evaluate the relative benefits of RFG with and without oxygenates and with various amounts and types of oxygenates. For these reasons, some discussion of the reliability of the models and their attendant uncertainties is in order.

Both the Complex and Predictive Models are based on statistical analyses of a large number of tests and the data used to develop both models are similar. Nevertheless, substantial differences exist. Some of these differences make comparison between the models cumbersome. For example, the Complex Model yields mobile emissions from a given RFG in units of milligrams per mile and the Predictive Model yields the percent reduction in the emissions from a given RFG blend relative to the so-called California Phase 2 reference fuel. There are also more-substantive differences. Probably most glaring of these is the fact that the Predictive Model ignores evaporative emissions. There are also differences of a more-subtle nature—e.g., the Predictive Model adopts a linear relationship between NO_x emissions and olefin content, and the Complex Model includes a linear and a quadratic term. As illustrated below, these differences can produce significant discrepancies between the results of the two models.

Figure 5-1 compares the Predictive and Complex Models' calculated reductions of NO_x and VOC emissions for four illustrative RFG formulations. (A more detailed discussion of how various RFG formulations fare using the Predictive and Complex Models is presented in Chapter 7.) Note that in all four cases and for both models, the use of RFG is predicted to lead to substantial emissions reductions relative to the federal baseline fuel (see Table 5-6). This is perhaps not surprising because RFG is intended to perform better than baseline fuels. However, this does not have to be the case a priori, and, in fact, there is evidence that an increase in emissions can result in some instances (Weaver and Chan 1997).

Inspection of Figure 5-1 indicates a good deal of consistency between the two models. For example, both models predict substantial benefits from the use of low sulfur fuel (i.e., ~90% reduction in sulfur relative to baseline fuel). (Indeed it appears that the use of low sulfur fuel will be critical to meeting the Phase II RFG requirements for VOC and NO_x emissions.) On the other hand, the models tend to diverge in their simulations of the effects of oxygenates. In the case of NO_x, the Complex Model shows a slight increase in emissions with the use of oxygenates, but little difference between the moderately and highly oxygenated fuels. However, the Predictive Model produces a varying effect, with little change with a moderate amount of oxygen and an increase in emissions with a high amount of oxygen. It is likely that this difference arises from the aforementioned different formulas used by the models to represent the effect of oxygenates on NO_x emissions. In other words, while both models are based on similar data, their different sets of covariates and associated parameters have apparently generated small, but non-negligible inconsistencies between model results. In the case of VOC emissions, the differences in the model results are far more substantial. For example, the Predictive Model yields greater emissions reductions as the oxygen content is increased, while the Complex Model predicts decreasing emissions reductions with the use of fuel with high amounts of oxygen.

The discrepancies illustrated in Figure 5-1 point to the possible existence of specific problems with one or both of the models. There are also some more-general concerns that need to be borne in mind. The use of the Complex and Predictive Models requires a substantial extrapolation of measured emissions from a sample set of motor vehicles operating under controlled test conditions to real-world emissions from a fleet of motor vehicles using one or more RFG blends. As discussed in Chap-

A. NO$_x$ Emissions

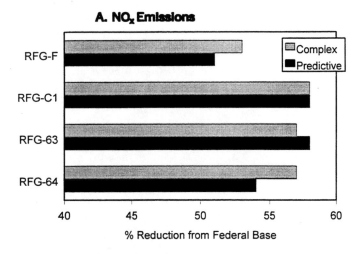

B. VOC Emissions

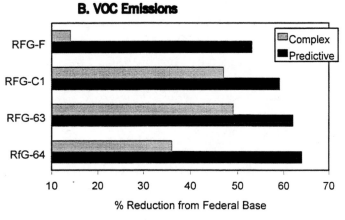

FIGURE 5-1 Recent reductions in the mobile-source emissions from four illustrative RFG blends relative to the federal base fuel (see Table 6-1) as predicted by the EPA Complex Model and the California Predictive Model. The reductions were determined assuming emission rates from federal base fuel of 1340, 907, and 500 mg/mi for NO$_x$, exhaust VOC, and evaporative VOC, respectively, and emissions from California Phase 2 reference fuel of 569 or NO$_x$ and exhaust VOC, respectively. Abbreviations: F, low aromatics; C1, low sulfur; 63, low sulfur plus moderate oxygen using MTBE; 64, low sulfur plus high oxygen using ethanol.

ter 4, there are myriad factors that can affect real-world vehicle emissions and confound attempts to produce a mobile-source-emissions model using statistics and regression models. Moreover, the characterization of the relationship between emissions measured in a controlled, testing program and those resulting from on-road driving remains a scientific and technological challenge. A potential source of error in both models arises from their treatment of high-emitting vehicles. As discussed in Chapter 6, a large portion of motor-vehicle emissions come from high-emitting vehicles. However, the emissions from these vehicles are likely to be quite variable and thus difficult to characterize through sampling a small subset of the total population. All the above issues will tend to limit our ability to use these models to assess the benefits of oxygenated RFGs. See Chapter 7 for additional discussion of results obtained from the Complex and Predictive models.

SPECIFICATION FLEXIBILITY AND DOWNSTREAM CONTROL IN FEDERAL PHASE II RFG

In the year 2000, Phase I RFG blends sold under the federal RFG program in the nine severe nonattainment areas and all present and future opt-in areas will be replaced by Phase II RFG. The targets for fuel content and exhaust-emissions reductions relative to conventional 1990 baseline gasoline are summarized in Table 5-6. In light of the focus of this report, it is relevant to note the Phase II requirement for a minimum oxygen content of 2% by weight.

The rules for meeting the requirements for Phase II are codified in 40 CFR 80.41 (e-f) and have been amended by subsequent action by removal of the per-gallon minimum NO_x reduction requirements for refiners using the pooled-averaging method. Other limits vary by whether an area lies in a northern- or southern-tier state. A refiner may select whether to meet product performance requirements on a per-gallon or pooled-average basis, as under California regulations. For example, with respect to benzene, if the former option is chosen, no single gallon of gasoline produced can contain more than 1% benzene by volume. If the latter is chosen, the pooled sample of all a refiner's RFG sold in nonattainment areas must average no more than 0.95%, meaning at least some of the gasoline must have less than the per-gallon maximum. Similarly, the oxygen content of an RFG sold by a refiner opting for the pooled-average method must average at least 2.1% by weight at any time, with no individual gallon falling below 1.5% but not exceeding 2.7% in winter if VOC-control requirements are in place.

There remains a concern about potential abuse of the process of adding oxygenate to gasoline downstream of a refinery. This practice, called "splash blending," involves mechanical mixing of finished gasoline or gasoline blending stock having front-end volatility set at a typical warm-season value (RVP of 7 to 8 psi) with a liquid oxygenate (such as ethanol). Splash blending, unlike refinery-performed match blending that renormalizes product output to the required properties of an RFG, can change the proportional constituents of a gasoline by diluting (replacing) their mass and volumetric share in each gallon. It also has the potential to increase the quantity of the total fuel that evaporates from vehicles if the fuel's resulting RVP is significantly higher. EPA sought to obviate this possibility by requiring the type of oxygenate that can be added be stipulated at the refinery and thus maintain RVP integrity. It also assures that even in the "worst case," with respect to volumetric displacement of benzene and other aromatics by an oxygenate (i.e., about 6% ethanol by volume in an ethanol-gasoline blend), Complex-Model content limits can be maintained by blending-stock planning at the refinery. EPA has instituted enforcement procedures to assure correct blending stock labeling, and the entire process for maintaining downstream RVP control is documented in the February 16, 1994, rulemaking on RFG standards (EPA 1994).

The possibility of an increase in the volatility of gasoline after leaving the refinery is expected to be low. Because refiners are held liable for the performance of their gasolines tested during EPA's in situ sample audits, most refiners now blend oxygen into summer RFG at the refinery (adding it in a controlled process to base gasoline at very low RVP, e.g., 6.7 psi or less). This is done to ensure that it matches Phase I property specifications. Because formulation stringency will increase for Phase II gasoline, this practice is likely to persist. Thus, splash blending should become a nonissue as applied to RFG formulation and sale during the ozone season. In fact, the demonstrated consistency of refining practice year-round has prompted EPA to remove the distinction between gasolines designated as "oxygenated fuels program reformulated gasoline" (OPRG) and those designated as non-OPRG, effective November 6, 1997 (EPA 1997b).

A related issue has to do with the fungibility of the gasoline supply (i.e., different blends of gasoline that comply with RFG requirements can be mixed freely in the distribution system as far downstream as the vehicle's tank and the resulting mixtures themselves comply with the requirements). The RVP of an ethanol blend can increase slightly if the volumetric share of the ethanol falls to a value between 5% and 10%. Thus mixing of Phase II RFGs with and without ethanol could lead to an in-use blend that does not meet Phase II RFG requirements. Recognizing that nonlinearities in the relationships between specific fuel properties and emissions could give rise in the gasoline distribution chain to a mixture of fuels that independently meet RVP specification but in combination violate it, EPA conducted extensive parametric variation testing within the Complex Model. Its conclusion was that use of various RFG blends within an area, would not give rise to scenarios in which application of the Complex Model showed nonconformity with specified emissions-performance requirements. (EPA 1994, pp. 7731-7732)

MODELING EVAPORATIVE VOC EMISSIONS FROM RFG FOR SIP DEVELOPMENT

Although EPA requires that the Complex Model be used to certify the properties of RFG, a meteorologically driven air-quality model is specified to derive the mobile-source emissions from vehicles using RFG for the purpose of assessing the air-quality benefits of the RFG program and demonstrating attainment of air-quality standards. In this method, the current version of the MOBILE emissions-factor algorithm is to be used as the basis for determination of the mass-emissions rate for exhaust and evaporative VOCs, CO and NO_x from highway vehicles that is appropriate to the local climatological regime and type of gasoline sold. Air-quality regulatory and planning organizations do not directly use the Complex Model in their forecasts, but may, for sophisticated air analyses, apply the Complex Model gasoline-property results obtained from refiners for the gasoline sold locally.

For purposes of complying with planning requirements for attainment of the ambient ozone standard, the values of the key variables needed for computation of evaporative VOC emissions in MOBILE (ambient temperature and gasoline RVP) should accurately represent the average conditions for the ozone "design-value" day. These are the conditions observed on the day that the regulated maximum ozone concentration—the datum from which ambient concentration reduction requirements is computed for purposes of SIP (State Implementation Plan) commitments—was recorded. MOBILE computes separate emissions factors for four nonexhaust components of nonmethane hydrocarbons (hot soak plus resting, diurnal, running, and refueling losses).

The Complex Model's determination of nonexhaust (evaporative) emissions of Phase II RFG also involves four separate computations. However, in each case, RVP (and distillation temperatures) are used to characterize these emissions (i.e., temperature is specifically not included as an independent variable) (40 CFR 80.45(c)(2-3)). The results of these four computations are summed to yield the nonexhaust component of the overall VOC-emissions-performance equation. The temperature conditions input to the Complex Model were based on average temperatures observed on ozone exceedance days which were estimated by EPA to be 72-92°F for the northern United States and 68-95°F for the southern United States. Thus, the Complex Model is broadly representative of high ozone conditions in the areas where RFG is sold. Accounting for the effects of variations in temperature on program implementations would involve considerations that are outside the scope of this study, such as the possible nonuniformity of RFG certification standards and possible complications with the distribution of RFG.

The absence of any temperature dependency in evaporative emissions computations in the Complex Model has raised concern that the model might assign too low a value to the nonexhaust VOC component of the RFG compliance calculation. At the very least it is quite possible that the VOC emissions derived from the Complex Model to certify fuels will be inconsistent with the emissions derived from MOBILE and used by regulatory agencies in the development of their SIPs. However, in preparing their SIPs, states can use the MOBILE model to estimate RFG effects on evaporative emissions by using more accurate local temperatures.

Is this discrepancy important? Certainly, as discussed in Chapter 4, high ambient temperature (and the magnitude of daily temperature rise) plays a role in the quantity of evaporative VOC emissions produced, and it is possible that current emissions-certification procedures underestimate the contribution of hot soaks to total evaporative emissions. Moreover, with low-volatility fuels such as RFGs, it appears that RVP differences, other things being equal, still dominate differences in total evaporative emissions for most relevant urban ozone nonattainment cases given current on-board emissions controls. Another concern is that refiners are limited, by current requirements, in their ability to craft fuels with lower total reactivity, if they chose to do so, because they might exceed exhaust plus evaporative mass-emissions targets. It is also the case, as alluded to above, that regression models inevitably introduce smoothing and other simplifying approximations that might be inappropriate in specific nonattainment areas, especially if they were developed from a data base different from that used to build any of the other regulatory models associated with the ozone-compliance process.

SUMMARY

Gasoline has been reformulated to adjust its basic properties for various reasons over a very long period of time, before, in fact, air quality became a major issue. The relatively recent emphasis on the control of ozone precursor emissions and toxic emissions has prompted a new and comprehensive gasoline reformulation strategy. This strategy involves: (1) reduction in summer volatility (expressed as RVP); (2) reduction in reactive gasoline components (e.g., olefins) during the summer to reduce the ozone-forming potential of motor-vehicle emissions; (3) reduction in benzene and other aromatic content of gasolines year-round, and (4) addition of oxygenates as a means to help control emissions and to maintain octane rating using nontoxic constituents. The first three of these are formally included in the federal and California reformulated gasoline programs.

The adoption and use of the Complex Model and Predictive Model have been driven by a need for establishment of a level playing field for all refiners, as well as an easy and inexpensive fuel certification procedure that allows mixing of different batches thereby facilitating fuel distribution. The models appear to meet those needs. However, the methods used in those models to predict the in-use performance of gasolines reformulated to meet the criteria of the reformulated gasoline programs, are based on results from large and diverse, but nonetheless limited, data bases. They might not accurately represent what actually occurs in specific nonattainment areas, especially where a high summer-temperature rise produces relatively high evaporative VOC emissions. They might even deny refiners the ability to formulate fuels that could be more beneficial on the basis of atmospheric reactivity—an issue that is addressed in Chapter 7.

6

The Effects of Reformulated Gasoline
On Ozone and Its Precursors

THE ABILITY TO distinguish the air-quality benefits of one reformulated gasoline (RFG) blend from that of another depends, to a substantial degree, on the overall magnitude of the effect of RFG on air quality. If the RFG effect is large, then the effect of two blends of RFG might be quite discernible. If on the other hand, the RFG has a lesser effect on air quality, it is likely to be very difficult to identify which of two RFG blends is preferable from an air-quality point of view, let alone to reliably quantify these effects. As a prelude to Chapter 7, in which an attempt is made to quantify and compare the ozone-forming potential of eight different RFG blends, this chapter assesses available information on the overall impact of the RFG program on ozone and its precursors as deduced from measurements.

The steps taken in the approach to make that determination are illustrated in the "Decision Tree" depicted in Figure 6-1. The chain of inference proceeds from the tests of the emissions from a limited sample of motor vehicles in the laboratory to determinations of the influences of the use of RFGs in light-duty vehicles (LDVs) on air quality. The figure also indicates the types of findings at each step along this chain, from considerations of the currently available and published observations to the reduction of ozone concentrations and other air-quality issues. The sequence of questions addressed are listed below.

131

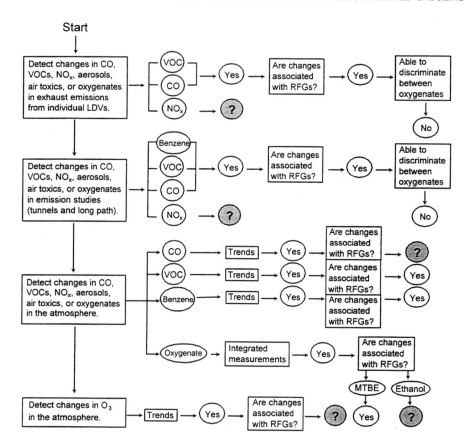

FIGURE 6-1 The Decision Tree illustrates the steps taken in an effort to quantify and compare the ozone-forming potential of various RFG blends. The figure indicates the types of findings at each step that resulted from the committee's considerations of the currently available observations that are pertinent to the reduction of ozone concentrations and other air-quality issues. When comparing different RFG blends, such as a blend containing ethanol versus a blend containing MTBE, it is desirable to account for as many differences as possible between the RFG blends.

•   What changes in motor-vehicle exhaust emissions of VOCs, NO_x, CO, or air toxics are observed in laboratory tests when RFGs are used?

•   Have the changes in emissions from RFGs indicated by laboratory studies been observed in emissions studies using tunnels and remote sensing of tailpipe exhaust?

- Are there data to support meaningful analysis of atmospheric data to determine the effect of RFGs?
- Have changes in the concentrations of air toxics or oxygenates been observed in the atmosphere and can these changes be related to the use of RFGs?
- Have changes in the concentrations of CO been observed in the atmosphere and can these changes be related to the use of RFGs?
- Have changes in the concentrations of ozone been observed in the atmosphere and can these changes be attributed to the use of RFGs?

This analysis proceeds from the information concerning the measurements of exhaust and evaporative emissions from individual vehicles to the observation of the effect of those emissions on atmospheric composition. When comparing two RFG blends, it is desirable to account for as many differences as possible between the RFG blends.

WHAT CHANGES IN MOTOR-VEHICLE EXHAUST EMISSIONS OF VOCs, NO$_x$, CO, OR AIR TOXICS ARE OBSERVED IN LABORATORY TESTS WHEN RFGs ARE USED?

Probably the most extensive single data set on the emissions of motor vehicles using RFG blends is that compiled from the Auto/Oil Air Quality Improvement Research Program (AQIRP).[1] This study included over 3,000 emissions tests. In Phase I of AQIRP, different sets of 26 reformulated fuels and 2 reference gasolines were tested in fleets composed of 20 then-current (1989) LDVs (cars and light-duty trucks) and 14 older vehicles (1983-1985). Further, two methanol blends (10% and 85% methanol in gasoline) and one industry-average fuel were tested in 19 flexible-fueled and 5 variable-fuel passenger vehicles. In Phase II of AQIRP, fuels were prepared in several sets, or matrices, to study the effects of individual fuel properties: (1) the composition set tested the effects of aromatic content, olefin content, T_{90} (temperature at which 90% of mass of the fuel has evaporated), T_{50} (temperature at which 50%

---

[1]The complete set of data for all experiments is available in reports and on CD ROM from the Coordinating Research Council, 219 Perimeter Center Parkway, Suite 400, Atlanta, GA 30346.

has evaporated), and the addition of methyl tert-butyl ether (MTBE));
(2) the RVP-oxygenate set tested the effects of Reid vapor pressure
(RVP), as well as the addition of ethanol, ethyl tert-butyl ether (ETBE),
and MTBE; (3) the methanol set tested various methanol-gasoline mixtures; and (4) the sulfur-series set tested effects of varying the sulfur
content of the fuel. The properties of the RFGs used in the AQIRP
compositional and sulfur tests (and those used in MTBE and ethanol
blends discussed in Chapter 7 of this report) are summarized in Table 6-
1.

Exhaust emissions were measured from the various vehicles as they
ran on a dynamometer under the Federal Test Procedure (FTP) protocol.
Gas chromatographic and high-performance liquid chromatographic
analyses of the exhaust emissions were made for all measurable components, including 140 structurally different hydrocarbons with from 1 to
12 carbon atoms, as well as ethers, methanol, ethanol, and 12 different
aldehydes and ketones. Samples of exhaust emissions were segregated
according to the point in the cycle of engine operation (cold start, hot
stabilized, hot start, and composite) to reconstruct the emissions inventories for various vehicular operating scenarios. For some fuel-vehicle
combinations, evaporative emissions were tested (modes of operation,
hot soak, diurnal, and running loss).

Emissions of Toxics

Many of the RFG blends used in the AQIRP studies showed significantly
lower total mass emissions of toxics than the industry-average gasoline.
This is illustrated in the comparisons shown in Figure 6-2 for industry-
average gasoline (A) and one of the RFG blends studied (C2). The
comparison is made for the older fleet, the current fleet, federal Tier 1
vehicles, and vehicles with "advanced technology." With the exception
of formaldehyde,[2] the RFG blends showed significantly lower toxic emissions for every class of vehicle when compared to emissions resulting
from the industry-average gasoline.

---

[2]Many RFG blends appear to result in an increase in formaldehyde exhaust
emissions. That is attributed to the presence of MTBE in the fuel, which can
generate formaldehyde during combustion.

TABLE 6-1 Properties of Some of the Research RFG Blends Used in AQIRP and California Studies

Code[a]	Composition Identifier[b]	Aromatics (vol %)	Oxygenates (vol %)[c]	Olefins (vol %)	T_{50} (°F)	T_{90} (°F)	RVP (psi)	Sulfur (ppm by wt)
AQIRP Phase I								
A	Industry average	32.0	0	9.2	218	330	8.7	339
B	Certified	29.9	0	4.6	220	309	8.7	119
C	AMot	43.8	15.4 (M)	3.3	213	288	8.7	284
D	amOT	20.7	0	22.3	218	357	8.5	316
E	AMOT	43.7	14.8 (M)	17.2	220	357	8.7	267
F*	amot	20.0	0	3.2	197	279	8.8	290
G	AmOt	44.3	0	17.4	214	286	8.8	317
H	aMOt	20.2	14.6 (M)	20.2	168	286	8.5	312
I	AmoT	42.9	0	4.1	239	353	8.9	261
J	aMoT	21.4	14.9 (M)	4.0	208	356	8.6	297
K	Amot	45.7	0	4.9	208	294	8.8	318
L	AmOT	47.8	0	17.7	236	357	8.5	266
M	aMOT	18.0	14.5 (M)	21.8	193	356	8.7	301
N	aMot	21.4	13.9 (M)	5.7	164	292	8.8	294
O	AMOt	46.7	14.6 (M)	19.3	204	283	8.6	288
P	amOt	20.3	0	18.3	190	284	8.5	333
Q	amoT	21.5	0	4.8	234	357	8.6	310
R	AMoT	46.0	15.2 (M)	4.0	225	354	8.4	279
S		21.2	0	3.8	199	280	8.0	297
T		18.1	9.7 (E)	3.6	174	276	9.8	246
U		19.1	9.7 (E)	3.1	171	278	9.6	278

Table 6-1 (Continued)

Composition Code[a]	Identifier[b]	Aromatics (vol %)	Oxygenates (vol %)[c]	Olefins (vol %)	T_{50} (°F)	T_{90} (°F)	RVP (psi)	Sulfur (ppm by wt)
AQIRP Phase I (continued)								
MM		22.2	14.8 (M)	5.4	167	289	8.0	345
AQIRP Phase II								
C1		22.7	0	4.6	208	297	6.9	38
C2		25.4	11.2 (M)	4.1	202	293	6.8	31
Y4		24.9	10.9 (M)	1.2	201	298	9.1	44
Y5		24.3	11.1 (M)	1.3	200	299	9.0	138
Y6		24.6	10.7 (M)	1.1	200	297	8.9	258
Y7		24.9	10.6 (M)	1.1	201	299	8.8	350
Y8		24.6	10.7 (M)	1.0	201	300	8.8	443
B2		26.7	0	2.5	220	318	8.9	49
Y2		26.1	0	2.3	220	316	8.8	466
California Studies								
63		23.4	11.6 (M)[d]	5.0	196	296	6.8	32
64		23.3	11.2 (E)[d]	4.8	188	297	7.8	34

[a] Fuel mixtures A-R are the compositional matrices for RFGs used in AQIRP Phase I; Y4-Y8 are sulfur matrices (with MTBE) from AQIRP Phase II; B2 and Y2 are from sulfur-varied fuels used in AQIRP Phase I with no added MTBE.

[b] Composition indicator: A/a, high/low aromatics; M/m, high/low MTBE; O/o, high/low olefins; T/t, high/low T_{90}.

[c] Oxygenates added are indicated with letters: MTBE (M), ethanol (E).

[d] For these two fuels, the oxygenate composition is given in mass %.

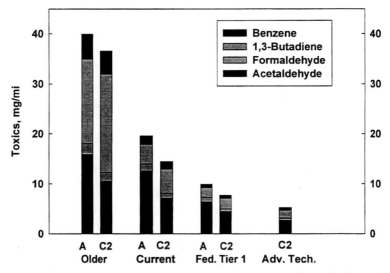

FIGURE 6-2 Comparison of the mass of exhaust toxics: acetaldehyde, formaldehyde, 1,3 butadiene, and benzene (mg/mi) from the industry-average fuel (A) and an RFG (C2), using the FTP composite. Chemicals are displayed from top to bottom as follows: acetaldehyde, formaldehyde, 1,3-butadiene, and benzene. On the x-axis, the results are divided into those for older, current, federal Tier-1-control, and advanced-technology cars. Source: Adapted from AQIRP Technical Bulletin No.17, 1995.

Emissions of VOCs

The specific and total reactivity (using the MIR scale) of VOCs in exhaust, evaporative (i.e., diurnal and hot soak), and running-loss emissions from current-fleet vehicles using several of the AQIRP-tested reformulated gasolines as well as the industry-average gasoline are shown in Figure 6-3. Speciation and reactivity data on exhaust emissions were obtained from Hochhauser et al. (1992); data on evaporative emissions and running losses were obtained from Burns et al. (1992). In the case of each type of emission, the ordering of the fuels has been adjusted to show the progression of emissions from the lowest-emitting fuel to the highest-emitting fuel. In viewing these figures, it should also be borne in mind that the ozone-forming potential of VOC emissions is determined by the total mass of the emissions as well as the reactivity of the species that are emitted. The relative contribution of each of these factors can be inferred by comparing the specific and total reactivities of the emissions because the specific reactivity is a measure of the amount of ozone formed per unit mass of VOC emitted and the total reactivity is the product of the specific reactivity and the mass of VOC (and CO) emitted per mile traveled (see Table 3-9). Finally, it should noted that in addition to emissions data for current fleet vehicles, AQIRP data exists for emissions from older fleet vehicles. Although the older-fleet data differ somewhat from that of the current fleet (e.g., the ordering of the fuels with increasing reactivity), the basic conclusions concerning the nature and magnitude of the emissions reductions that might be obtained from RFG do not.

Inspection of Figure 6-3 indicates that rather substantial changes in the reactivity of VOC emissions can result from variations in gasoline formulation. In the case of exhaust emissions for example, the specific reactivities of the fuels tested vary by a factor of 1.4, and total reactivities by a factor of about 2 (Figure 6-3A). The variability in the reactivities of diurnal and hot soak emissions are of a similar magnitude, although the ordering of the fuels changes significantly (Figure 6-3 B and C). By comparison, the range of running-loss reactivities is considerably larger (i.e., factor of 2 variability in specific reactivity and a factor of almost 70 in total reactivity) (Figure 6-3D).[3] However, the maximum reduction in the reactivity of the VOC missions obtained by switching from the industry-average formulation to the most favorable of the RFGs tested is, in each case, considerably smaller. For exhaust, diurnal, and hot soak emissions, the reduction in specific and total reactivity from the industry average is about 25% or less. In the case of running losses, the reduction is more substantial; i.e., a factor of about 2 for the total reactivity.

Of course the most important parameter to consider here is the composite reactivity of all the LDV emissions; i.e., the reactivity obtained from the gases emitted by all exhaust, evaporative, and other loss processes. An example of such composite reactivities is given in Figure 6-4. In this case, composite, specific reactivities were calculated for each fuel using the AQIRP measurements of exhaust emissions (weighted for all cycles of operation), evaporative emissions, running losses, resting losses, and refueling losses from LDVs using the EMFAC-7E emissions model and the measured vapor pressures of the fuels. The relative

---

[3]According to Burns et al. (1992), running loss emissions were measured at less than 0.2 g/test on all but two vehicles in each fleet. In the vehicles which had higher running losses, differences could be seen between fuels, but fuel effects could not be determined because of the limited data and its variability.

139

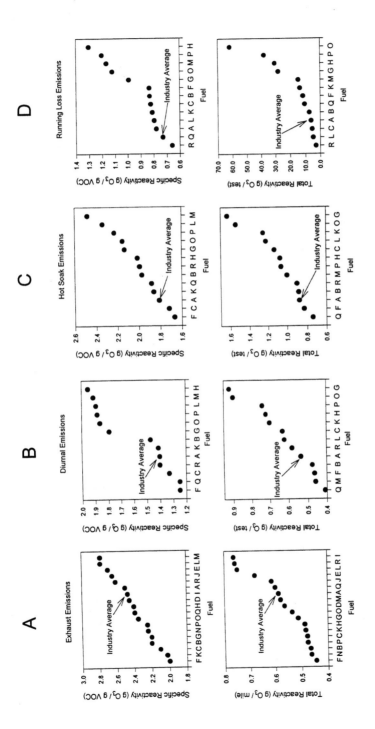

Figure 6-3 AQIRP current fleet vehicles using various RFGs (see Table 6-1) and industry-average fuel. Reactivities are expressed using the MIR scale. (A) Exhaust emissions; (B) diurnal evaporative emissions; (C) hot-soak evaporative emissions; and (D) running-loss emissions. Source: Burns et al. 1992 and Hochhauser et al. 1992.

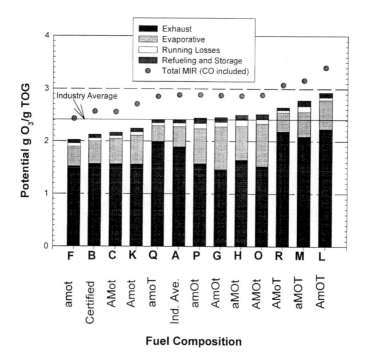

Fuel Composition

FIGURE 6-4 Comparison of the specific reactivity (potential g O_3/g VOC for the total VOC emissions) with the contribution of using industry-average fuel A and various RFGs. (TOG (total organic gas) is considered to be interchangeable with VOC.) The properties of the fuels and the compositional abbreviations shown on the x-axis are described in Table 6-1. Emissions are displayed in the bars from top to bottom as follows: refueling and storage, running losses, evaporative, and exhaust. For data represented by circles, the mass of CO emissions is not included in the denominator of the specific reactivity values plotted. The addition of CO reflects the importance of a very low reactivity compound that is emitted along with the VOCs. Source: Adapted from AQIRP Technical Bulletin No. 12, 1993.

weighting of the various emissions to produce a composite emission was made to simulate the conditions present in Los Angeles, California in 1995. Here again we find substantial differences in the reactivities resulting from the fuels tested. The fuel range of reactivities from the least reactive fuel (F) to the most reactive fuel (L) is a factor of about 1.5. However, the reactivity resulting from the least reactive fuel is only about 20% less than that obtained with the use of the industry-average fuel.

Another interesting facet of the reactivities in Figure 6-4 relates to the role of CO. Note in the figure that the circle above each of the bars represents the specific reactivity for the appropriate fuel when the reactivity of the CO emissions is included. The average increase in the reactivities from the inclusion of CO is 18 ± 2%; the ordering of the fuels is also changed somewhat. These results clearly demonstrate the need to include CO emissions when assessing the ozone-forming potential of LDV emissions.

Exhaust Emissions of NO_x

The AQIRP data suggest that the effect of RFG on exhaust emissions of NO_x will vary depending upon the specific properties of the blend. For example, NO_x emissions were lowered by 6 ± 1.9%[4] by reducing olefin content from 20 to 5%, while reducing T_{90} from 390°F to 280°F increased NO_x emissions by 5 ± 2.4%, and the impact of lowering aromatic VOC content did not have a statistically significant effect (i.e., NO_x emissions were lowered by 2.1 ± 2.0%).

The effect of adding oxygenates to the fuel tended to produce a small increase in NO_x emissions. For example, increasing ethanol from 0 to 10% gave rise to a 5 ± 4.1% emissions increase. On the other hand, while adding 15% MTBE and 17% ETBE also resulted in an emissions increase, the increase was not statistically significant (i.e., 3.6 ± 5.4% for MTBE and 5.5 ± 6.4% for ETBE). The average of experiments with added oxygenates was a statistically significant increase of 4.8 ± 2.9%.

By far the largest decrease in NO_x emissions were achieved by lowering the sulfur content of the fuel. This effect is discussed in more detail in the next section.

Effect of Fuel Sulfur Content of RFGs on Exhaust Emissions

Dramatic changes in exhaust emissions of all ozone precursors (i.e., VOC, CO, and NO_x) were obtained from the sulfur set of AQIRP tests (see Table 6-1). In these tests, the fuel's sulfur content was varied while the

---

[4]All uncertainties are twice the standard deviations of the mean expressed as 2σ or 95% confidence levels.

aromatic and alkene VOCs and MTBE compositions, as well as the T_{90} values, were kept relatively constant (e.g., fuels Y4 to Y8). Figure 6-5 shows that the mass of hydrocarbons (HCs) and CO, and NO_x in the exhaust gases generally increase with increased sulfur in the fuel. In Figure 6-6, it can be seen that the mass (milligrams per mile) of the total toxics, benzene, 1,3-butadiene, and acetaldehyde tend to increase with increasing sulfur content of the fuel, and that of formaldehyde decreases somewhat.

Thus, the data from the Auto/Oil Study suggest a very clear effect on the emissions of most reactive organic species as well as NO_x, CO, and toxics with the use of low sulfur-containing fuels. In contrast, the effects of sulfur content of the fuel on engine-out emissions (i.e., no flow through the catalysts) were found to be very small. This suggests that the sulfur effect is related to a temporary decrease in catalyst efficiency, most likely because the sulfur reacts with and alters the catalyst surface. However, these effects appear to be largely reversible as sulfur in the fuel is decreased.

Effect of RVP on Emissions

Another major contributor to the reduction of LDV emissions is lowering a fuel's RVP, which significantly reduces evaporative VOC emissions. (Exhaust emissions are also reduced, to some extent, by compositional changes made to RFG blends to lower their RVP.)   Overall, lower RVP appears to be the major contributor to lowered VOC emissions resulting from the use of RFG. It is important to note that before implementation of the RFG program, reductions in RVP were mandated and likely led to a significant decrease in VOC emissions. However, appropriate monitoring networks were not in place at that time, and it has been difficult to quantify the impact of RVP reduction.

HAVE THE CHANGES IN EMISSIONS FROM RFG BLENDS INDICATED BY LABORATORY STUDIES BEEN OBSERVED IN EMISSION STUDIES USING TUNNELS AND REMOTE SENSING OF TAILPIPE EXHAUST?

The laboratory tests of the AQIRP indicate that RFG can result in significant decreases in the emissions of the ozone-forming precursors (reactive

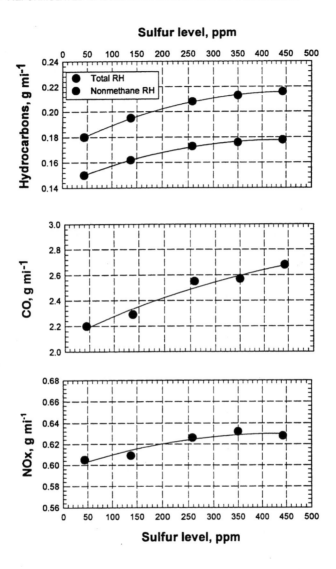

FIGURE 6-5 Tailpipe emissions (g/mi) for HCs, CO, and NO_x from current-fleet vehicles fueled with various RFGs that have very similar hydrocarbon compositions but contain different amounts of sulfur compounds (fuels Y4 to Y8 in Table 6-1). In contrast, the engine-out emissions show very little effect of the sulfur content of the fuel, consistent with the importance of sulfur-catalyst interactions that lower the effectiveness of the catalyst. For the plot of hydrocarbons versus sulfur level, the upper curve corresponds to total hydrocarbons (total RH) and the lower curve corresponds to nonmethane hydrocarbons. Source: AQIRP Technical Bulletin No. 8, 1992.

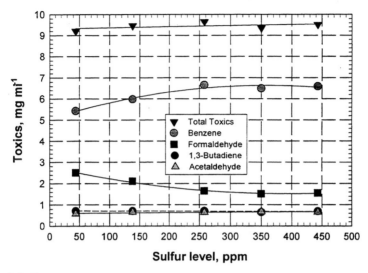

FIGURE 6-6 Tailpipe emissions (mg/mile) for the toxic compounds (acetaldehyde, formaldehyde, 1,3-butadiene, and benzene) from current-fleet vehicles fueled with various RFGs that have very similar hydrocarbon compositions but contain different amounts of sulfur compounds (fuels Y4 to Y8 in Table 6-1). Source: AQIRP Technical Bulletin No. 8, 1992.

VOC, NO$_x$, and CO), as well as toxics. In its second level of investigation, the committee focuses on whether the effects of RFG blends seen in these laboratory tests are also found in the emissions of motor vehicles operating under actual driving conditions. Emissions studies of this nature are typically carried out in two ways: (1) measuring aggregate emissions from motor vehicles within a tunnel; and (2) measuring tailpipe emissions of individual motor vehicles using remote sensing. It is important to recognize that these types of measurements do not provide a comprehensive measurement of emissions from motor vehicles. Tunnel measurements are conducted in a restricted environment, and as such, they are neither air-quality measurements nor emissions measurements. The emissions data from tunnel studies only measure exhaust plus evaporative running losses, are highly aggregated, and represent a snapshot of on-road emissions of a representative vehicle fleet for specialized driving conditions. Remote-sensing of tailpipe exhaust, on the other hand, largely measures exhaust emissions. Despite these limitations, tunnel studies and remote-sensing provide an important calibration point between automotive emissions tests (such as those carried out in the AQIRP) and studies that attempt to identify a signal from ambient measurements.

Tunnel Studies

A series of tunnel studies was conducted by Kirchstetter et al. (1996, 1997, 1999a,b) for the period from 1994 to 1997 when fuels in California changed from Phase 1 RFG to Phase 2 RFG. In 1994, measurements were performed in August and in October for vehicles operating on California Phase 1 RFG fuels with and without the addition of winter oxygenates, thus allowing for assessment of the effects of oxygenates on emissions. In each year, intense measurements were conducted for 10 days or more during late July and early August. The vehicle fleet, ambient conditions, driving conditions through the tunnel, the tunnel, and the ambient air quality are described in detail in Kirchstetter et al.(1999a). This research affords one of the best opportunities to examine the effects of various types of RFGs on emissions.

Vehicular emissions were measured in the Caldecott tunnel, a heavily used commuter tunnel that runs in the east-west direction through the Berkeley Hills near Berkeley, California, connecting Contra Costa County residential areas to San Francisco. The tunnel has three two-lane bores, and on weekdays, traffic through the central bore is switched from the downhill westbound direction in the morning to the uphill eastbound direction in the afternoon. The tunnel is about 0.7 mi (1.1 kilometer) long, has a nearly constant grade of +4.2% in the eastbound direction, and has fully transverse ventilation provided by adjustable pitch fans.

Sampling was conducted between 4:00 p.m. and 6:00 p.m. when vehicles were traveling in the uphill eastward direction. The nearest on-ramp providing access to the center bore of the tunnel is located more than 0.6 mi away, ensuring that all vehicles in the center bore were in the warmed-up mode. Vehicle counts per hour during the sampling period averaged approximately 4,200. The mean model year for the fleet driving through the tunnel was 1989.3 for the 1995 study, 1990.1 for the 1996 study, and 1990.9 for the 1997 study. Averages were slightly less than the median values. Approximate average fleet composition was 67% cars, 33% vans and sport-utility vehicles, and less than 0.3% heavy-duty trucks; however, light-duty trucks increased from 31% to 35% during the period from 1994 to 1997, with cars exhibiting a corresponding decline. Vehicles traveling through the tunnel were in the hot stabilized mode and averaged 37 mph. Average vehicle speed at the entrance was 32 mph and at the exit was 43 mph. Instrumented vehicular measurements performed during extensive drive-through in 1996 provided additional information about driving conditions. Heavy acceleration and stop-and-go driving were seldom observed. Most of the driving in the tunnel occurred within a small range of speeds and accelerations that is largely within the FTP domain.

Continuous measurements of CO, CO_2, and NO_x were made in the tunnel exhaust air at a location close to the exit. Background gas concentrations were determined by making measurements in the in-coming ventilation air. Concentrations of CO, NO_x, and VOCs were typically 25, 30, and 10 times higher in the tunnel air compared with background air. Two-hour integrated air samples for quantifying hydrocarbons and carbonyls were taken concurrently with the continuous measurements, and analyzed within 48 hr by gas chromatography and high-performance liquid chromatography.

The 1994 Caldecott Tunnel Studies

The 1994 studies of Kirchstetter et al. (1996) are described separately because they afford an opportunity to examine the effects of California Phase 1 gasoline with and without the addition of winter oxygenates. Average properties of gasoline used during various segments of the 1994 study are given in Table 6-2. Unfortunately, in addition to the changes in oxygen content, other fuel properties changed as well. For example, there is a small increase in both sulfur and RVP in the winter gasoline. Each of these tends to result in increased VOC emissions. The increased sulfur content also tends to increase CO and NO_x emissions (see Figure 6-5).

The data from the study suggest that the addition of oxygenates (in the form of MTBE) to the fuel during October appeared to lead to a reduction in CO and VOC emissions of $21 \pm 7\%$ and $18 \pm 10\%$, respectively. A similar reduction in CO emissions ($16 \pm 3\%$) was measured during the Colorado oxygenated fuels program (Bishop and Stedman 1990). NO_x emissions showed no change during the two sampling periods. In the case of toxics, formaldehyde emissions increased by $13 \pm 6\%$ and benzene emissions decreased by $25 \pm 17\%$, but no significant change on acetaldehyde emissions was observed.

The addition of MTBE also appeared to lead to changes to the relative abundances of individual VOCs and thus might have affected the reactivity of the emissions. However, analysis of the data indicated that

TABLE 6-2 Average Properties of California Bay Area Phase 1 RFG for August and October 1994

Fuel Property[a]	Sampling Period[b]	
	August 1994 (Low Oxygenates)	October 1994 (High Oxygenates)
Oxygen content (wt %)	0.3 ± 0.4	2.0 ± 0.2
Sulfur (ppm by wt)	54 ± 47	90 ± 53
Reid vapor pressure (psi)	7.2 ± 0.2	7.7 ± 0.3
Paraffins (vol %)	47 to 54	38 to 46
Aromatics (vol %)	34 to 43	26 to 35
Olefins (vol %)	0.4 to 7.3	4.3 to 13.4
Naphthenes (vol %)	2.9 to 10.4	4.1 to 9.6
Benzene (vol %)	1.7 to 5.1	1.0 to 3.6

[a]Gasoline composition was determined by the California Air Resources Board (CARB), and was based upon averaging 65 samples during the August period and 54 samples during the October period. On an oxygen weight basis, 80% of the oxygenate was MBTE and 20% was ethanol.

[b]Errors are reported as 1 σ (standard deviation) of the mean.

Source: Adapted from Kirchstetter et al. 1996.

the normalized VOC reactivity, using the MIR scale (see Chapter 3), did not change significantly from the low-oxygenate to the high-oxygenate period.

The 1994-1997 Studies

During the five sampling periods of this study, California gasoline changed composition from the 1994 summer and fall values indicated in Table 6-2 to California Phase 2 RFG. The evolution of the average summer gasoline properties during the study is summarized in Table 6-3. Emissions of all pollutants decreased by between 20% and 40% over the 1994 to 1997 study period (Kirchstetter et al. 1999a). However, attributing these changes to a specific cause such as an RFG blend is problematic because of the difficulty in separating the effects of fleet turnover from those of fuel changes. Using a statistical time-series analysis to separate these two effects, Kirchstetter et al. concluded that the effect of

TABLE 6-3 Summer (July and August) Gasoline Properties in California from 1994 to 1997

Year	RVP (psi)	API Gravity	Oxygen (wt %)	Sulfur (ppm wt)	Saturates (vol %)	Olefins (vol %)	Aromatics (vol %)	MTBE (vol %)	Benzene (vol %)	ASTM D-87			Density (kg/m³)
										T_{10}	T_{50}	T_{90}	
1994													
Avg.	7.4	54.4	0.51	131	57.4	7.9	31.9	2.7	1.56	136	214	334	761
sd	0.1	1.9	0.32	41	4.8	4.4	2.1	1.7	0.39	3	8	8	8
1995													
Avg.	7.4	54.7	0.21	81	56.5	8.8	33.7	1.0	1.54	136	218	341	760
sd	0.1	1.0	0.18	36	5.1	3.5	3.3	0.9	0.45	3	4	8	4
1996													
Avg.	7.0	58.9	1.96	16	62.6	3.3	23.5	10.7	0.42	138	199	300	743
sd	0.1	0.6	0.30	9	2.5	0.9	1.4	1.7	0.08	2	4	4	2
1997													
Avg.	7.1	59.5	1.57	12	65.4	3.4	22.7	8.2	0.43	138	200	299	741
sd	0.1	1.3	0.60	11	3.7	1.2	1.4	3.7	0.05	2	3	6	5

Abbreviations: Avg., average; sd, standard deviation.
Source: Adapted from Kirchstetter et al. 1999a.

an RFG was greater on VOCs than on NO$_x$. No significant change was observed in acetaldehyde emissions, whereas the effect of RFG blends on benzene was estimated to be a 30% to 40% reduction, and on formaldehyde, a 10% increase.

Kirchstetter et al. (1999b) attempted to characterize evaporative emissions that are affected by gasoline vapor pressure (i.e., those due to refueling, running loss, and diurnal evaporation). Bay Area gasoline was analyzed to determine composition, and headspace vapor composition was estimated using the Wagner equation (see Reid et al. 1987). The individual compound vapor pressures were determined from the vapor pressure of the pure species, its mole fraction, and activity. Combining that information with the emissions data from the tunnel, the change to California Phase 1 RFG was estimated to cause a 13% vapor-phase reduction in evaporative emissions, and the change to Phase 2 caused a further 9% reduction, giving rise to a net reduction of evaporative emissions from California RFGs of 20%. (Normalized reactivity of liquid gasoline and headspace vapors decreased by 23% and 19%, respectively.) Combining that result with those for exhaust emissions indicated that the ozone-forming potential (measured as total reactivity by the MIR scale) of all on-road emissions decreased by 8% or less as a result of RFG blends. The total reactivity decrease was less than that of evaporative mass emissions because of increased weight fractions of highly reactive iso-butene and formaldehyde in the exhausts (from the combustion of MTBE).

Collectively, the Caldecott tunnel studies suggested that there were significant reductions in the motor-vehicle emissions of all pollutants (except formaldehyde) between 1994 and 1997. These decreases, summarized in Table 6-4, are attributable to a combination of the use of RFG and fleet turnover effects, with RFG most likely making a significant contribution. However, some caution should be exercised before using these results to characterize the overall effect of RFGs on motor-vehicle emissions. As noted above, any number of factors can have a significant effect on emissions from motor vehicles (e.g., age, stop-and-go driving, and cold-start conditions) that are minimally represented in the Caldecott tunnel.

Finally, Kirchstetter et al. observed high concentrations of ethene and acetylene in the tunnel, which are indicative of reduced catalytic-converter activity that results in high-emitting vehicles. Because decreases in sulfur concentration in gasoline have little effect on such high-emitting vehicles, one interpretation of the study results suggests that high emitters, such as older-technology vehicles or vehicles with faulty

TABLE 6-4 Decrease in Emissions 1994-1997 Inferred from the Caldecott Tunnel Studies[a]

Emission	% Decrease
CO	31 ± 5
Nonmethane VOC	43 ± 8
NO_x	18 ± 4

[a]A fraction of these decreases are attributable to RFG and a fraction to fleet turnover.

catalytic converters, might be responsible for a disproportionate share of the VOC emissions in the tunnel. As discussed below, a similar conclusion was reached by Beaton et al. (1995) using on-road remote sensors in urban locations in California.

Remote-Sensing Studies

Remote sensing has been used to infer the amount of ozone precursors (CO, NO and total VOCs) in exhaust emissions from individual in-service LDVs relative to the emission of CO_2 in the exhaust (Bishop and Stedman 1989, 1990; Bishop et al. 1989; Guenther et al. 1991, Zhang et al. 1993, 1996a,b, Stedman et al. 1994, 1997, Butler et al. 1994). In this technique, light at specific wavelengths in the infrared (IR) or ultraviolet (UV) spectrum is passed through the exhaust plumes of passing motor vehicles. The measurement and analysis, which is based on the amount of light absorbed by the compounds contained in the exhaust, have been shown to quantitatively determine CO emissions in the exhausts to within $\pm 5\%$ and VOC emissions to within $\pm 15\%$ (Lawson et al. 1990; Stephens and Cadle 1991; Ashbaugh et al. 1992).

In typical studies, the technique is deployed at frequently traveled roadways, such as freeway entrances and exits. Used that way, the technique measures exhaust emissions under nominal roadway operation. The technique has been deployed at a variety of locations in the United States and abroad. Provision can be made to record the identity of individual vehicles to determine the emissions demographics of the vehicle fleet. The technique has been extensively used to establish the trends in emissions as a function of vehicular age, to monitor the effectiveness of vehicular maintenance and inspection programs, and to measure the effect of the addition of oxygenated compounds to fuel to reduce emissions of CO and VOCs.

Remote-sensing measurements show that, in every model year from pre-1971 to 1991 for the given fuel supply (Zhang et al. 1993; Beaton et al. 1995), there has been a steady reduction in exhaust emissions of VOCs and CO (see Figure 6-7 for data on VOC emissions). The measurements also indicate that, although there has been a steady decline in exhaust emissions in the more recent models, the exhaust emissions for each model year are dominated by a relatively small number of high emitters (i.e., the vehicles in the fifth quintile in Figure 6-7). The results show that, for each model year, properly maintained vehicles provide only a small contribution to the emissions from that model year compared with poorly maintained or malfunctioning vehicles.

The remote-sensing technique has been used extensively in Denver to study the efficacy of the use of oxygenated fuels to reduce CO exhaust emissions from LDVs. In Denver, CO pollution is most severe during the winter and the Colorado program specifically targeted reductions in winter CO emissions from LDVs by the addition of oxygenated compounds to the gasoline sold in Colorado. In the first Denver study, the CO emissions were measured from approximately 60,000 vehicles at a freeway on-ramp during and after the oxygenated fuel season. Bishop and Stedman (1989) reported a 6 ± 2.5% reduction in CO attributable to the use of oxygenated fuel containing 2.0 wt % oxygen. In a second Denver study, Bishop and Stedman (1990) analyzed vehicular emissions from more than 117,000 vehicles at two Denver locations (a freeway on- and off-ramp) before, during and after a winter season when oxygenated fuels were mandated (November 1988 through February 1989). They reported a 16 ± 3% decrease in CO emissions from the use of oxygenated fuel at 2.0% oxygen.

A followup study using this technique was carried out in Denver to determine the effectiveness of the 1991-1992 winter Colorado oxygenated fuels program (PRC 1992). Based on the results, the percentage reduction of CO emissions was nearly the same for all vehicles, and most of the reduction in CO emissions attributed to oxygenated-fuel use were from the highest-emitting vehicles (Figures 6-8 and 6-9). Even though a small portion of the vehicles tested were high emitters, those vehicles contributed a substantial portion of the CO emissions. The study indicated a comparable result for the reduction in exhaust emission of VOCs.

An important finding of these remote-sensing measurements was that most of the overall CO and VOC emissions and the reductions in these emissions from the use of RFGs are associated with emissions from high emitters and, more specifically, from vehicles with malfunctioning emissions controls. The studies (Bishop and Stedman 1989, 1990, 1995;

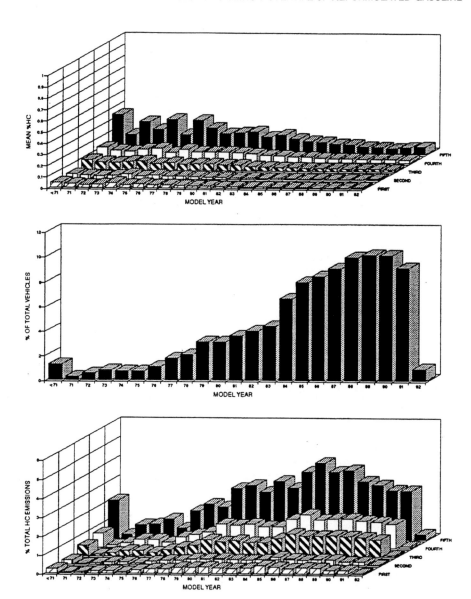

FIGURE 6-7 Location-specific data from the Denver area using remote sensing. Part A shows emission factors by model year divided into five groups (quintiles) in ascending order of emissions. Part B shows the vehicle age distribution of the measured fleet. Part C is the product of data from Parts A and B; percentage of total HC (or VOC) emissions is shown for each quintile of each model year. Source: Zhang et al. 1993. Reprinted with permission from *Environmental Science and Technology*, copyright 1993, American Chemical Society.

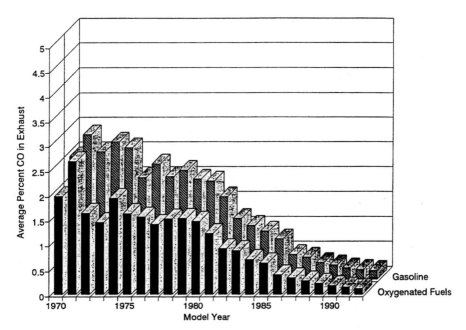

FIGURE 6-8 CO emissions by model year of motor vehicle as recorded at a specific location in the Denver area. Source: PRC 1992.

Bishop et al. 1989; Guenther et al. 1991; PRC 1992; Zhang et al. 1994, 1996a,b; Beaton et al. 1995; Stedman et al. 1997) find that the addition of oxygenated fuels reduces the exhaust emissions of CO and VOCs by approximately 20% for all model years in LDVs with the largest emissions. For example, Beaton et al. (1995) placed on-road remote sensors of exhaust CO and VOC emissions at various urban locations in California. They found that 7% of the vehicles accounted for more than 50% of CO emissions and 10% of the vehicles accounted for more than 50% of the VOC emissions. This group probably involves LDVs that were not well maintained or have otherwise improperly functioning emissions control systems. Because this finding was independent of the model year, it implies that a relatively small percentage of vehicles with the highest exhaust emissions will be the principle sources of the exhaust emissions of CO and VOCs and that the use of oxygenated fuels in those vehicles will be of the greatest benefit by reducing the exhaust emissions of CO and VOCs by approximately 20%.

The remote sensing methods, by virtue of their deployment, observe LDVs under "cold-start" conditions, and emissions estimates based on the method will fail to account for the cold-start fraction of the total emis-

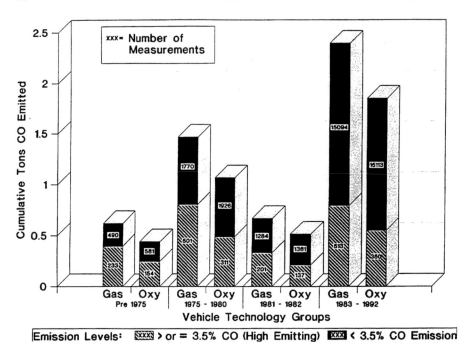

FIGURE 6-9 CO emissions contribution from high-emitting vehicles (hatched) and lower-emitting vehicles (black) by vehicle-technology grouping. High emitters are defined as having emissions with more than 3.5% CO; lower-emitting vehicles have emissions with less than 3.5% CO. The number on each bar segment refers to the number of cars recorded in the sample of vehicles measured at a specific location in the Denver area. Source: PRC 1992.

sions. Because catalytic converters will operate at reduced effectiveness under cold-start conditions, the relative importance of high emitters under these conditions might be less. If cold-start emissions represent a sizable fraction of the total exhaust emissions, the benefits of oxygenated fuels may be more uniformly spread across the light-duty vehicle fleet.

ARE THERE DATA TO SUPPORT MEANINGFUL ANALYSIS OF ATMOSPHERIC DATA TO DETERMINE THE EFFECT OF RFGs?

The tunnel studies and remote-sensing measurements discussed in the previous section have provided useful information concerning the effects of RFG blends on the emissions of a variety of ozone precursors, including CO from LDVs. However, it is difficult to relate these snapshot assessments of LDV exhaust and running losses to the actual net effect of RFGs on air quality. To accomplish this, the use of atmospheric measurements as an assessment must be examined.

The attribution of the trends of ambient ozone concentrations and those of its precursors to specific control policies is complicated by the presence of confounding influences such as meteorological fluctuations (NRC 1991; Rao et al. 1992, 1998; Cox and Chu 1993; Milanchus et al. 1998). Therefore, answering questions such as "what portion of the change in ambient ozone concentrations can reasonably be attributed to a particular emissions control policy?" requires the existence of high-quality, long-term concentration data gathered in a carefully designed network. Therefore, it is critical that a spatially well-designed monitoring network be in place to measure precursors as soon as possible. When time series of ozone and precursor data covering the pre- and post-implementation time periods are available for the regions where the control program is in effect and where it is not in effect, one can apply space-time analyses and change-point detection techniques as suggested by Rao et al. (1998), Hogrefe et al. (1998), and Zurbenko et al. (1996) to observe the effects of the emission control strategy on ambient pollutant levels.

During the last 30 years, there have been extensive data sets acquired from integrated field measurements and the estimation of long-term trends from those measurements. However, these measurements were not aimed at determining specifically the effectiveness of particular air-quality regulations. The measurements were aimed at assessing the reductions in concentrations of criteria pollutants or to determine the processes or sources of the primary emissions that limit these reductions. In the case of ozone, which is formed by secondary photochemical reactions, these measurements were not designed to determine the alteration of ozone concentrations that results from the RFGs. Unfortunately, when the planning of integrated field measurements or monitoring fails to include directed observations to document a particular aspect of air-quality regulations, it is generally not possible to isolate these effects from the data that have been acquired for other purposes. For these reasons, the ability to discern an "RFG signal" in the ambient data sets is quite limited. At this time, researchers are only able to even attempt such an analysis for a limited set of relevant species: RFG oxygenates, toxics, CO, and ozone. Rao et al. (1998) concluded that a goal of trend assessment should be to isolate and characterize long-term (greater than 1-year concentrations of pollutants) information based on multi-variate analyses of ambient weather, climate, and emissions. All long-term variation should be considered without regard to a particular trend model (e.g., linear, step, or ramp).

HAVE CHANGES IN THE CONCENTRATIONS OF AIR TOXICS OR OXYGENATES BEEN OBSERVED IN THE ATMOSPHERE AND CAN THESE CHANGES BE RELATED TO THE USE OF RFG?

Benzene is both an ozone precursor and an air toxic and, as a result, regulations have specifically targeted its reduction. LDV emissions of benzene are derived directly from benzene and from higher aromatics in the fuel. The RFG programs, with their prescribed reductions in benzene and other aromatics (see Table 5-3), are intended to reduce ambient benzene concentrations. Although reductions have been observed in the atmospheric concentrations of benzene over the past several years (EPA 1998), the observations are not capable of attributing these reductions to a particular control strategy or to differentiate between different oxygenates used in fuels. Because oxygenated compounds were added to RFGs specifically to replace benzene and other aromatic compounds, it is reasonable to assume that at least part of the observed reduction in ambient concentrations is associated with the reduction in vehicular emissions as a source. To date, although reductions are observed at many locations in various VOC concentrations including larger aromatic compounds, the trends are not sufficiently consistent to draw definite conclusions.

Both MTBE and ethanol have been observed to be present in the atmosphere. These compounds can serve as ozone precursors, but because their atmospheric reactivity is low, they are not expected to be as effective as more-reactive VOCs in generating ozone in urban environments. However, like benzene and CO, they might be more effective in ozone formation farther downwind of the source of their emission. Because the only identified use for MTBE is as a motor-fuel additive, it is reasonable to assume that its presence in the atmosphere is associated with the emissions from LDVs using fuels with an MTBE additive. In this connection, MTBE could serve as an important tracer to determine the influence of its addition to motor fuel on the other compounds of interest.

By contrast, ethanol has many natural and anthropogenic sources. To date, no analysis has yet been carried out to determine if or how much of the burden of ethanol in the atmosphere is associated with its use as a fuel additive.

HAVE CHANGES IN THE CONCENTRATIONS OF CO BEEN OBSERVED IN THE ATMOSPHERE AND CAN THESE CHANGES BE RELATED TO THE USE OF RFGs?

Because motor vehicles are the primary source for CO, the U.S. Environmental Protection Agency (EPA) required that urban areas classified as nonattainment for CO use oxygenated fuels in gasoline-fueled engines during the winter season beginning in 1992. CO is primarily a winter problem because low surface temperatures limit the dispersion of the pollutant and enhance its emissions from cold engines. As outlined in Table 5-4, oxygenated fuels in most CO nonattainment areas are blended to contain a minimum of 2.7% oxygen by weight.

The Oxygenated Fuels Program has now been in effect for at least five winters in several different metropolitan areas, a time interval that might be long enough to begin an assessment of whether or not this program has or has not been effective. In fact, recently, a number of researchers have attempted to assess the impact of fuels on ambient CO concentrations (Mannino and Etzel 1996; Cook et al. 1997; Dolislager 1997; Whitten et al. 1997). Those studies have generally concluded that the oxygenated fuels program has resulted in a discernable downward trend in ambient CO concentrations. However, in the committee's view, the studies are not conclusive. The Oxygenated Fuels Program was initiated in the midst of other control programs and technological improvements designed to lower CO emissions. Colorado, for example, places restrictions on both wood-burning stoves and driving times when CO concentrations are likely to be high. Most likely, such other programs and improvements have had some downward effects on CO emissions. Discerning the portion of the downward CO trend in an area that is specifically attributable to oxygenated-fuel use is a challenging problem. Another problem arises from inhomogeneities and discontinuities in the way in which CO data are reported. During the 1990s, the reporting of CO data in the United States was changed from rounding to the nearest 1 ppm to the nearest 0.1 ppm. Such discontinuities can produce an artifact in a trend analysis that confounds identification of an impact of a control program.

   To illustrate such problems, CO data from areas using oxygenated fuels were analyzed. It is important to note that this analysis is not intended to be a comprehensive assessment of the relationship between oxygenated fuel and ambient CO but just an illustration of the difficulties such analyses can encounter. The regions analyzed are broadly the states of New York, New Jersey, and Connecticut (NY/NJ/CT); Colorado; and California. Within the NY/NJ/CT region, the New York City Metropolitan Statistical Area (NYCMSA), implemented its program during the winter of 1992-1993. Colorado implemented a statewide oxygenated fuels program in 1988, and California during the winter of 1992-1993.

   All data used in the analysis presented here were obtained from the EPA's Aerometric Information Retrieval System (AIRS). The monitoring sites considered are listed in Table 6-5. There are 2 sites in Connecticut, 10 sites in New Jersey, 8 sites in New York, 16 sites in California, and 10 sites in Colorado. With the exception of California, these sites were chosen because of the length of their CO time-series records. The sites in California are the same as those used by Dolislager (1997), and are sites which have reasonably complete records that include violations of the 8-hr standard for CO during1990-1993.

   There are some problems with the raw hourly data because of the way in which the lowest values (detection limits) were reported. For the northeastern-states sites and Colorado sites, retaining only the daily maxima of 1-hr concentrations eliminates this problem. However, for the sites in California, the hourly data were first rounded to the nearest part per million prior to extracting the maxima. Rather than examining the reported daily maxima, their logarithms were examined for this report to help stabilize the variability due to seasonal variation.

   The various physical processes reflected in each of the CO time-series were separated into three components that contribute independently to the overall trend. These components are a short-term component (attributable to fluctuations in weather and day-to-day emissions); a seasonal-variation component (attributable to the Earth's rotation around the Sun); and a long-term component (attributable to secular, or lasting, changes in climate or emissions). It is this last component that is most important here, because the effect of control policies must be manifested in this component.

   Table 6-5 shows the amount contributed by each component to the total variance of the data at each monitoring site. In all three of the regions studied here, the short-term component contributes the most to the total variance, especially in the NY/NJ/CT region. The contribution of

TABLE 6-5 AIRS Data from Selected Sites in California, Colorado, and Northeastern States (1980-1997)

AIRS ID	Station Location		Percent Contribution to the Total Variance			Implementation Date (yymmdd)	% Improvement (Δ)	2σ-Level for Δ (%)[a]
			Short-term	Seasonal	Long-term			
090010004	Bridgeport	CT	74.4	6.6	15.5	921101	1.4	5.2
090010020	Stamford	CT	49.3	9.4	36	921101	-22.2	12.2
340035001	Hackensack	NJ	75.6	5.4	16.1	921101	7.6	4.8
340051001	Burlington	NJ	60.6	12.7	21.4	N/A	11.2	5.0
340070003	Camden	NJ	76.3	10.5	8.9	N/A	-7.1	5.5
340071001	(Not in a city)	NJ	63.8	14.5	15.6	N/A	-5.8	8.5
340171002	Jersey City	NJ	74.5	1.6	22.4	921101	9.4	5.6
340232003	Perth Amboy	NJ	62.8	3.8	30.1	921101	-3.0	7.8
340252001	Freehold	NJ	63.9	5.3	27.7	921101	-7.8	3.9
340270003	Morristown	NJ	64.7	3.6	28.7	921101	4.6	3.7
340292001	Toms River	NJ	72.9	5.4	17.6	N/A	1.8	5.7
340390003	Elizabeth	NJ	79.0	2.6	16.6	921101	7.9	6.3
360290005	Buffalo	NY	81.8	2.5	13.4	N/A	-7.1	3.2
360290016	Buffalo	NY	77.4	2.5	17.6	N/A	-8.9	4.8
360551004	Rochester	NY	84.7	2.9	10	N/A	-11.6	8.2
360556001	Rochester	NY	67.5	5.1	23.8	N/A	6.4	5.8
360590005	(Not in a city)	NY	78.6	6.0	12.2	921101	9.3	3.7

Table 6-5 *(Continued)*

AIRS ID	Station Location		Percent Contribution to the Total Variance			Implementation Date (yymmdd)	% Improvement (Δ)	2σ-Level for Δ (%)[a]
			Short-term	Seasonal	Long-term			
360610062	New York City	NY	49.6	3.3	43.9	921101	-4.2	10.6
360632006	Niagara Falls	NY	80.4	2.9	13.6	N/A	5.7	6.5
360930003	Schenectady	NY	73.5	11.2	10.4	N/A	0.1	11.8
060190008	Fresno	CA	49.1	37.4	1.3	921101	-4.3	13.0
060371002	Burbank	CA	49.3	32.7	8.3	921101	-1.7	5.2
060371103	Los Angeles	CA	63.7	26.1	2.2	921101	1.3	3.0
060371201	Reseda	CA	57.5	26.7	7.4	921101	-7.5	7.4
060371301	Lynwood	CA	44.2	41.8	1.4	921101	1.8	5.4
060372005	Pasadena	CA	57.6	29.1	3.8	921101	-3.9	5.3
060375001	Hawthorne	CA	51.1	37.4	1.5	921101	1.6	4.7
060590001	Anaheim	CA	45.9	35.4	7.7	921101	-3.0	9.1
060591003	Costa Mesa	CA	49.0	37.8	1.6	921101	4.7	8.6
060595001	La Habra	CA	49.4	36.1	4.4	921101	12.2	6.1
060670006	Sacramento	CA	67.0	21.6	3.7	921101	1.0	5.4
060670010	Sacramento	CA	61.6	26.7	1.6	921101	2.4	7.1
060771002	Stockton	CA	65.4	24.2	2.1	921101	11.2	5.2
060850004	San Jose	CA	48.9	35.7	4.1	921101	24.3	5.6

060950004	Vallejo	CA	59.0	29.5	1.2	921101	3.1	4.3
060990005	Modesto	CA	59.2	27.8	2.8	921101	16.0	5.9
080050002	Littleton	CO	64.2	19.0	9	880101	-3.0	11.6
080131001	Boulder	CO	56.0	18.2	19.4	880101	9.4	13.2
080310002	Denver	CO	55.3	14.6	25	880101	32.5	6.8
080310013	Denver	CO	61.1	16.0	17.4	880101	4.4	3.7
080310014	Denver	CO	58.1	22.7	11.8	880101	5.7	5.1
080410004	Colorado Springs	CO	64.0	15.0	15.4	880101	5.3	6.8
080410006	Colorado Springs	CO	72.3	12.0	10.9	880101	11.8	5.8
080590002	Arvada	CO	60.3	19.7	13.4	880101	23.7	7.0
080691004	Fort Collins	CO	61.7	22.6	9.1	880101	3.7	12.2
081230007	Greeley	CO	56.4	29.4	5.2	880101	15.6	4.2

[a]A positive value corresponds to a decrease in CO levels; a negative value corresponds to an increase.

the long-term component is large in NY/NJ/CT and Colorado, but very small in California.

The behavior of the CO time-series at Riverside, California, is presented in Figure 6-10. (Note that Riverside is not included in Table 6-5.) A change in the detection limits and resolution or data-reporting practices around 1994 is apparent from an inspection of the lower values shown in Figure 6-10A. The strongest decline in CO levels has occurred since 1987 (see Figure 6-10D). It should be noted that California introduced Phase 2 RFG in 1996 and winter oxygenates in 1992. The presence of strong downward trends in CO throughout the time series in the post-1987 period complicates evaluation of mid-series changes to regulatory policy. Examination of CO concentrations before and after implementation of the oxygenated fuels program might very well indicate a decrease in CO, but this decrease may be indicative of the overall downward trend that began well prior to the implementation of the program as opposed to the program itself.

To discern the contribution of oxygenated fuels to a trend such as that depicted in Figure 6-10, an analytical approach is needed that attempts to identify an abrupt "break" or change in the trend line at the time the program was first implemented. One such approach uses a linear regression on the long-term component (i.e., trend) for the period prior to the program implementation. That linear trend, prevailing prior to implementation of the program, is removed from the long-term component of the entire time series. Linear trends are then estimated for the detrended data for the pre- and post-implementation time periods. (By definition, the slope and intercept of the trend for the pre-implementation time period are zero.) The change in the intercept (Δ) in the pre-implementation time period at the date of fuels program implementation is an estimate of the percent change in CO concentrations attributable to that program. A confidence level (2σ) for the change, Δ, is also computed. Δ is positive (negative) for a decrease (increase), i.e., improvement (deterioration), in CO levels.

Values of Δ for each site included in this study are listed in Table 6-5 along with their respective 2σ confidence intervals. When Δ is greater than 2σ, the value derived for Δ is statistically significant at the 95% confidence level; when Δ is smaller than 2σ, the effects of the oxygenated fuels program on CO at that site cannot be discerned reliably from the data.

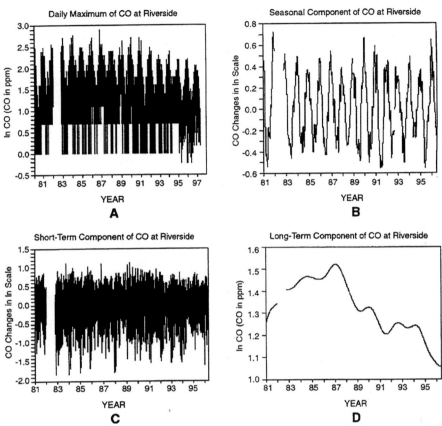

FIGURE 6-10 Daily maxima of CO concentrations at Riverside, CA, from 1980 to 1997 (A). Three components of the overall trend are seasonal (B), short term (C), and long term (D).

Examination of the Δ and 2σ values in Table 6-5 reveals varied results for the sites.[5] Most sites had positive Δ values (indicative of a benefit from the oxygenated fuels program). However, a substantial fraction (14 out of 46) of the sites had negative Δ values, and for many of the sites (23 out of 46) the Δ values were not significant at the 2σ (or

---

[5]Very similar results were obtained when the date of program implementation was shifted by ±6-month increments or when the analysis was restricted to data gathered during the months of October to February—the period when oxygenated fuels are used.

95%) confidence interval. Thus, while this analysis suggests that the oxygenated fuels program probably has had some small ameliorative effect on CO concentrations, its impact does not appear to be spatially uniform and in many cases is too small to discern with a high degree of statistical confidence.

A very similar conclusion was reached in a report of the National Science and Technology Council (NSTC 1997). The NSTC report reviewed various studies relating to the ambient air-quality effects of oxygenated fuels. It concluded that CO concentrations in urban areas have been decreasing at a rate of 2.8% per year for the last 10 years. This decrease is attributable primarily to EPA-mandated motor-vehicle emissions standards and improved vehicular emissions control technology. However, the NSTC report concluded that the benefits of oxygenated fuels on ambient air quality in cold climate areas could not be confirmed. (See Anderson et al. (1994) for additional information on the influence of oxygenated fuels on ambient CO.)

HAVE CHANGES IN THE CONCENTRATIONS OF OZONE BEEN OBSERVED IN THE ATMOSPHERE AND CAN THESE CHANGES BE RELATED TO THE USE OF RFGs?

Assessing the effects of RFG on ambient ozone air quality involves challenges similar to those discussed above for CO. For example, Larsen and Brisby (1998) attempted to assess the effect of California's cleaner-burning gasoline program on ozone concentrations. In that study, for the Sacramento, South Coast, and San Francisco Bay areas, Larsen and Brisby reported ozone decreases of 14%, 17%, and 4%, respectively. However, the contribution of cleaner-burning gasoline to this decrease is uncertain because of the presence of many other ongoing ozone-mitigation efforts. To address this problem, Larsen and Brisby assumed that the contribution of the cleaner-burning fuels program to the overall ozone decrease was proportional to the estimated percent reduction in the precursor emission inventory resulting from the program. Thus, even though the Larsen and Brisby study was based on ambient ozone concentrations, the attribution of a portion of the observed ozone decrease to the use of cleaner-burning gasoline was derived from an emission inventory and does not constitute empirical verification of program effectiveness.

To further illustrate some of the difficulties with applying trend analysis to ambient ozone data, consider the log-transformed ozone concentrations from Riverside, California, presented in Figure 6-11.[6] As in the CO analysis, the data are decomposed into its long-term, seasonal, and short-term components. Because the information from the moving-average filter (Zurbenko et al. 1996) used here is not reliable at the beginning and at the end of the time-series, data for the first and last years are not included in these figures. At this site, the long-term, seasonal, and short-term components contribute about 2%, 63%, and 34%, respectively, to the total variance of the ozone data.

To examine whether the use of RFGs in California had an impact on ambient ozone concentrations, data during the 1980-1997 period from several locations in the Los Angeles Air Basin of California were also analyzed. As was the case for the CO analysis in the previous section, an overall downward trend in ozone over the past 15-year period is evident in the long-term component at Riverside (Figure 6-11D). Between 1981 and 1996, ozone has decreased by about 30% at Riverside; the largest decrease of about 20% in ozone concentrations occurred between 1989 and 1993. Ozone then increased slightly in 1994, and then decreased again in 1995.

Whereas the oxygenated fuels program was implemented in California in 1992, the RFG program was implemented in 1996. Figure 6-11 indicates the presence of a strong downward trend in ozone before these programs were implemented. Unfortunately, data for the time period after the RFG program was implemented are not yet available for this type of analysis to clearly discern the impact of this control strategy on ozone air quality. For example, if an abrupt change of 10% in the middle of ozone time-series data illustrated in Figure 6-11 were introduced, it would contribute only about 0.5% to the total variance. This illustrates that the detection of any abrupt change of the order of 10% or less and its attribution to a specific control of an emission is a formidable task.

These results demonstrate the difficulty in linking a particular emissions-control policy to a change in ozone concentrations. Clearly, the problem of assessing the effectiveness of a particular air-pollution control program requires further development.

---

[6]The rationale for using the log-scale for ozone was discussed by Rao et al. (1997).

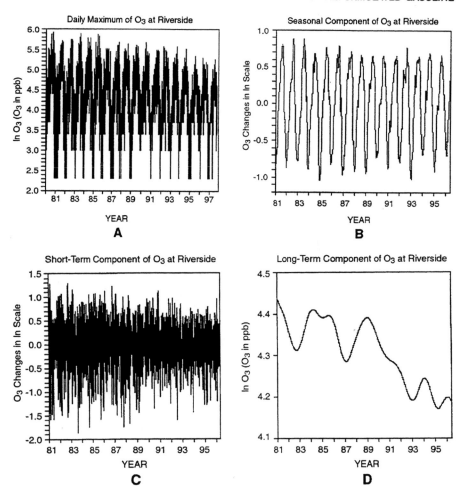

FIGURE 6-11 Daily maxima of ozone concentrations at Riverside, CA, from 1980 to 1997 (A). Three components of the overall trend are seasonal (B), short term (C), and long term (D).

DOCUMENTATION OF RFG EFFECTS IN A FUTURE OBSERVATIONAL PROGRAM

On January 1, 2000, federal Phase II reformulated gasoline (RFG) will be required in commercially available LDVs operating in areas classified as being in severe nonattainment of the National Ambient Air Quality Standard (NAAQS) for ozone. On the basis of estimates from the Complex Model, EPA expects that this action will result in reductions in both exhaust and evaporative emissions of VOCs and some air toxics from LDVs, as well as LDV exhaust emissions of CO and NO_x. It is further believed that these emissions reductions will help alleviate the severity of the ozone pollution in the severe nonattainment areas where the program is to be implemented, although, for the reasons discussed above, these effects are not expected to be large or even observable.

Will the projected air-quality benefits of Phase II of the federal RFG program be met? As with any regulatory program, the committee recommends that a complete and comprehensive RFG program should include—part and parcel—a plan for documenting the impact of the program and assessing to what extent the expected benefits are realized. The committee further recommends that this plan be organized around addressing a progression of three scientific questions[7] that attempt to document the effect of Phase II RFG on ozone precursor compounds and their ozone-forming potential. (Ideally, such a plan would include a fourth question that addresses the effect of the Phase II RFG on ozone concentrations. However as discussed above, it is unlikely that such a signal in ambient ozone concentrations could be discernible given the relatively large variability in ozone, the myriad factors that affect ozone concentrations, and the rather small overall impact RFG is projected to have on ozone.) The three questions recommended here for consideration are briefly discussed below.

Question 1: Do in-use Phase II RFG blends decrease the emissions from LDVs?

This first question can be addressed in much the same way that the potential air-quality benefits of RFG were initially assessed in studies such as the AQIRP and California Ethanol Testing Program (see Chapter 7). Representative vehicles can be selected and then subjected to emissions tests using dynamometers, etc. In this case, however, actual, in-use Phase II RFG would be used instead of prospective RFG formulations. Fungibility issues, such as that related to in-tank blending of RFGs, could then, in principle, be directly tested and assessed.

---

[7]These questions tend to mirror the progression of questions included in the Decision Tree in Figure 6-1.

Question 2: Are changes in emissions resulting from the use of Phase II RFG blends observable under driving conditions?

Although measurements of LDV emissions in a laboratory setting are informative, they do not necessarily represent the emissions of LDVs in operation under actual driving conditions. Confirmation that laboratory-measured emission reductions also occur on the road can be obtained through tunnel studies and remote sensing of tailpipe emissions. As noted earlier in this chapter, these measurements characterize LDV emissions under a limited set of conditions and, as such, do not comprehensively quantify LDV emissions. Nevertheless, they do provide a real-world test of the emissions and as such are an important step in linking laboratory-measured LDV emissions to an ambient concentrations signal.

Question 3: Are changes in emissions resulting from the use of Phase II RFG blends observable as a signal in the ambient concentrations of ozone precursor compounds?

Establishing the connection between changes in LDV emissions and the ambient concentrations of the compounds contained in those emissions is a more-formidable task. The most-straightforward approach for accomplishing such a task is through the use of time-series analyses of a long-term record of ambient concentrations of VOC, CO, and NO_x to isolate a signal that can be associated with Phase II RFG. However, this approach presents a variety of challenging problems. The time-series record must encompass a period significantly before as well as after initiation of Phase II RFG and the data set must include highly accurate and precise measurements. Even under those circumstances, identification of a shift in the time series of the quantity of interest due to RFG can be obscured by other transient factors (e.g., meteorological variations or implementation of other emissions control programs). Therefore, there is a need to develop and evaluate techniques for detecting ambient effects of a control program separately from the effects of meteorological variability.

For those reasons, it is recommended that an alternative approach be taken to document the effect of Phase II RFG usage on ambient precursor concentrations. This alternate approach would be to use measurements of various tracers in conjunction with measurements of

VOC, CO, and NO_x to (1) characterize the contributions of LDV emissions to the concentrations of ozone-precursor compounds; (2)estimate the ozone-forming potential of these compounds through the application of various observation-based methods (e.g., Cardelino and Chameides 1995); and (3) document the change in this contribution that can be attributed to the use of RFG. Tracer species that would be useful in this regard include those that could be used to identify LDVs emissions (e.g., acetylene for LDV exhaust), as well as those that could serve as a fingerprint of emissions from LDVs using RFG (e.g., MTBE). These measurements would ideally be made in a variety of locations within and surrounding each severe nonattainment area to document effects occurring on regional scales as well as local or urban scales. Especially important in this regard would be the enhancement of monitoring capabilities in rural areas of the United States.

SUMMARY

The first investigation in this chapter focused on determining if changes in ozone precursors (NO_x or VOCs, CO, air toxics, and oxygenates) have been observed in the emissions studies done on individual vehicles tested under controlled conditions in the laboratory. The most comprehensive study undertaken to date of the effects of varying gasoline properties, the Auto/Oil Air Quality Improvement Research Program (1989-1995), indicated that substantial ozone-precursor emissions reduction benefit should be achieved by RFG. Decreases in the ozone-forming potential (as measured by the MIR scale) of emissions from LDVs of as much as 20% appear to be possible. The most dramatic effects on ozone-precursor exhaust emissions seen in the various gasoline compositional matrices studied were those due to lowering the fuel's RVP and the amount of sulfur-containing compounds. Only slight reductions, less than 10%, in the CO and VOC emissions can be ascribed to the addition of either MTBE or ethanol.

The second investigation focused on determining if changes in NO_x or VOCs, CO, air toxics, and oxygenates, have been observed in the emissions studies done in tunnels or from remote sensing of exhaust. From a qualitative point of view, these studies appear to be consistent with the laboratory tests. Reductions were observed in the LDV emissions of NO_x, VOCs, CO, and various toxics, and they appear to be at least partially attributable to the introduction of RFGs. Formaldehyde emissions were found to increase—most likely from the combustion of MTBE. These studies also indicated that high-emitting vehicles are responsible for a disproportionate share of the VOC and CO emissions. The tunnel studies and remote-sensing measurements also indicated that the addition of oxygenates to fuel substantially reduced the emissions of CO and VOCs from these high emitting vehicles, perhaps because these high-emitters are operating with faulty or nonfunctioning catalytic converters. However, the data from these studies could not be used to discern the relative air-quality benefits of fuels using MTBE or ethanol.

The third and final investigation sought to identify RFG effects in the atmosphere by analyzing ambient data. Such an undertaking is easily confounded by competing and offsetting interferences (e.g., meteorological variations and the existence of other contemporary control programs), and statistically significant trends specifically attributable to the RFG program could not be identified. Several areas of the country have seen significant improvements in air quality, including reductions in ambient CO and ozone concentrations. In the case of CO, it appears that some portion of the decrease can be attributed to the addition of oxygenates to fuels but the magnitude of the oxygenate effect is not spatially uniform and in some areas is too small to discern with statistical confidence. In the case of ozone, it is not clear if any portion of the concentration decrease can be directly associated with the addition of oxygenated compounds to motor fuel or the development and use of RFG.

Thus, it would appear that RFGs have an impact on ozone-precursor emissions from LDVs by reducing both the mass and ozone-forming potential of these emissions. However, discerning a statistically significant effect of RFGs on ambient ozone concentrations has thus far proven to be quite difficult. This is most likely because ambient ozone concentrations tend to be quite variable from year to year and the RFG program is but one of a multitude of ozone-mitigation programs underway in the nation whose impact on ozone is of a similar or larger magnitude. Thus, air-quality models—which are themselves subject to significant uncertainty—present the only avenue for estimating the magnitude of the effect of RFG on ozone concentrations. As described in Text Box 6-1, simulations using these models indicate that the overall reduction in ozone from the implementation of the RFG program is likely to be a few percent. This finding should not be interpreted to mean that RFG use is

TEXT BOX 6-1   Model Predicted Effects of RFG
On Ground-Level Ozone

Laboratory tests and tunnel studies suggest that the use of RFGs in LDVs lowers the ozone-forming potential (as measured by the MIR scale) of an individual vehicle's emissions using an RFG blend with the lowest MIR by about 20% (see Figure 6-4). Yet, analysis of ambient data is unable to identify a discernible impact on ground-level ozone concentrations. Does that indicate an inconsistency or gap in our understanding of the processes that lead to the formation and accumulation of ozone pollution? Not necessarily. In the first place, ozone concentrations generally do not respond in a linear fashion to decreases in VOCs (see discussion in Chapter 2). Moreover, emissions from LDVs represent only a fraction of the total VOC emissions in an airshed. Thus, it might be expected that the effect on ambient ozone of a ~20% decrease in the reactivity of motor-vehicle emissions would be considerably less than 20%.

A more quantitative assessment of the probable impact of RFGs on ozone can be made using air-quality models. One could ask, Are changes in emissions resulting from the use of RFG blends observable in air-quality models, and has the performance of those models been evaluated? A version of the gridded Urban Airshed Model was exercised as part of the AQIRP study to do just such an assessment (AQIRP 1997a). In this study, the Urban Airshed Model was used to simulate ozone concentrations when different RFG fuels were used for conditions typical of Los Angeles, New York, and Chicago-Milwaukee. Simulations were first carried out for a historical ozone episode in each metropolitan area (Los Angeles, August 26-28, 1987; New York, July 9-11, 1988; and Chicago-Milwaukee, June 24-28, 1991). RFG effects on ozone were then estimated using the same meteorological conditions that occurred during the historical episode and emissions projections for 2000 and 2010 that included the emissions reductions for motor vehicles predicted by the data from the Auto/Oil study. Table 6-6 lists the predicted change in peak ozone for each simulation for changes in T_{50}, T_{90}, and sulfur content of the fuels. As might be expected, lowering these fuel properties does in fact lead to a decrease in peak ozone concentrations. However, the ozone decrease is quite small—about 1 part per billion by volume (ppb) or less—although in many cases still statistically significant.

(Continued on next page)

(Continued from previous page)

An independent model assessment of the impact of the federal RFG program was carried out by the New York State Department of Environmental Conservation using the emissions inventory prepared by the Ozone Transport Assessment Group (OTAG). The study involved a regional-scale application of the Urban Airshed Model (UAM-V) with a model domain covering much of the eastern third of the continent. The regions where the RFG program was implemented during 1995 is presented in Figure 6-12. A comparison of model simulations of a multi-day ozone episode during July 7-18, 1995, with and without the RFG program indicates ozone decreases up to 3 ppb over Chicago, Lake Michigan, and along the northeastern corridor (see Figure 6-13). Of course it should be recognized that air-quality models simulations are themselves uncertain because of the uncertainties in both the algorithms (e.g., the chemical mechanisms) and the input data (e.g., the emission inventories) used to run the models. Even recognizing these uncertainties, it seems unlikely that the RFG program could result in ozone decreases of more than 10 ppb. For example, even if the mobile source emissions used in the model simulations were underestimated by a factor of 2, the maximum ozone decrease would probably be less than 10% at most because peak ozone concentrations generally respond nonlinearly to changes in ozone precursor concentration.

Thus, model simulations predict that RFG has a beneficial effect of a few percent on overall ozone concentrations. It is therefore not surprising that discerning an RFG-signal in the ambient ozone data has proven to be difficult. It also suggests that it will be difficult to discern the impacts of two RFG blends with subtle differences in their properties. This issue is addressed as a case study in Chapter 7.

ineffective. As noted earlier, reduction of RVP in gasoline prior to the RFG program is thought to have had a significant air-quality benefit. As discussed in the next chapter, such a reduction size limits the ability to document the benefits of RFGs and to reliably distinguish between the ozone-forming potentials of different RFG blends.

TABLE 6-6 Predicted Effects on Peak Hourly Ozone Concentrations Expected Due to Changes in Certain Fuel Composition Variables in Three Cities As Estimated by Using the Urban Air Shed Model[a]

City, Year, Episode Day[c]	Change in Fuel Variable[b]		
	T_{50}[d](215°F to 185°F)	T_{90}(325°F to 280°F)	Sulfur (320 to 35 ppm)
	Change in Peak Ozone (ppb) from That of the Historical Episode		
Los Angeles, August 28, 1987			
2000	-0.3 ± 0.3*	-0.9 ± 0.3*	
2010	-0.1 ± 0.2	-0.1 ± 0.2	
New York, July 11, 1988			
2000	-0.1 ± 0.1	-0.4 ± 0.1*	-0.4 ± 0.1*
2010	0.0 ± 0.1	0.1 ± 0.1	-0.4 ± 0.1*
Chicago–Milwaukee, June 26, 1991			
2000	-0.8 ± 0.7*	-1.2 ± 0.9*	0.0 ± 0.9
2010	-0.2 ± 0.7	-1.0 ± 0.8*	0.4 ± 0.8
Chicago–Milwaukee, June 27, 1991			
2000	-0.6 ± 0.5*	-0.9 ± 0.7*	-0.2 ± 0.7
2010	-0.1 ± 0.4	-0.5 ± 0.4*	0.1 ± 0.4
Chicago–Milwaukee, June 28, 1991			
2000	-0.3 ± 0.2*	-0.5 ± 0.3*	-0.2 ± 0.3
2010	-0.1 ± 0.2	-0.3 ± 0.2*	0.0 ± 0.2

[a]The predicted effects may not be reflective of the greatest change in gasoline composition such as changes from the late 1980s to when California Phase 2 RFG began to be used.

[b]Main effects are shown with 95% confidence intervals. An * denotes statistically significant effects.

[c]Data from the location and date that was used to establish meteorological conditions employed in each simulation.

[d]The effects of T_{50} on ozone may be underestimated because only the effects on emissions from lower exhaust emitters are included. The effect of T_{50} on emissions from higher emitters could not be estimated from the available data and are assumed to be zero.

Source: AQIRP Technical Bulletin No. 21, 1997a.

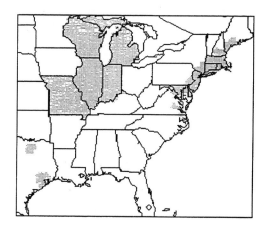

FIGURE 6-12   The areas where the RFG program was implemented during 1995.

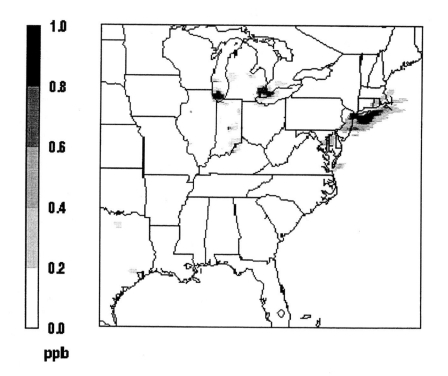

FIGURE 6-13   Maximum change in ozone from RFG as predicted by the UAM-V model for July 7-18, 1995 episode.

7

Using Ozone-Forming Potential to Evaluate the Relative Impacts of Reformulated Gasolines: A Case Study

As DISCUSSED IN Chapter 6, reformulation of gasoline has the potential to substantially reduce the light-duty motor-vehicle (LDV) mass emissions of VOCs, NO_x, and CO, as well as air toxics. Moreover, the emissions reductions resulting from the use of many of these formulations are sufficiently large to satisfy the requirements of the federal Phase II and California Phase 2 Reformulated Gasoline (RFG) programs. Thus, it is believed that the federal and California RFG programs will have a mitigating impact on ozone pollution, although various analyses suggest that the magnitude of the effect is not likely to be large (i.e., on the order of a few parts per billion) even if emissions from LDVs are underestimated by a factor of 2 or so.

This chapter turns to a more-subtle and more-difficult issue: namely, discerning the relative air-quality benefits of RFG blends using different amounts and types of oxygenated compounds. Because the mass of VOC emissions can be a misleading indicator of the ozone-forming potential of these emissions, the committee assessed the air-quality benefits of various RFG blends on the basis of the reactivity of these emissions as well as their mass. It should be noted at the outset, however, that this is a difficult task. Recall from Chapter 6, that the overall reduction in the reactivity of LDV emissions from the use of RFGs (over that from conventional gasoline) is at most about 20%. The variation in the reactivity of emissions arising from various RFGs that differ in relatively minor ways (e.g., in oxygen content) is likely to be substantially smaller. On the other hand, recall from Chapter 3 that the uncertainty in the reactivities of a composite set of VOCs arising from a single source, such as motor vehicles, is probably also generally about 20%. Thus, a major challenge in this analysis was determining whether the difference in the reactivities of LDV emissions derived for two or more RFGs is statistically significant. In the analysis presented here, the committee adopted the so-called "paired *t* test"[1] to make this determination.

In the sections that follow, a brief overview of the paired *t* test and its relationship to statistical uncertainty is provided. This methodology was applied to assess the statistical significance of differences in the LDV emissions arising from a subset of fuels studied by Auto/Oil Air Quality Improvement Research Program (AQIRP) and the California Air Resources Board (CARB). These fuels and their general properties are listed in Table 7-1 (and more detailed fuel properties are given in Table 6-1). Two approaches are used to estimate the LDV emissions from these fuels: one based solely on the experimental data arising from the emissions studies themselves, and the other using the Complex and Predictive Models. In order to assess the role of oxygenates and, more specifically, the relative roles of MTBE and ethanol, the subset of fuels included in this analysis was selected to provide a range of oxygen contents from 0 to 3.4% by weight (recall that the federal RFG program calls for a minimum oxygen content of 2% by weight), with this oxygen coming from MTBE or ethanol.

The subset of fuels used in this study were chosen to look for the effects of substituting MTBE by ethanol in otherwise closely similar fuels. Clearly, it would be preferable to use data on MTBE-containing and ethanol-containing fuels with the same fuel oxygen content or similar oxygenate volume percent, with all other chemical and physical properties (other than the presence of MTBE or ethanol) being the same. However,

---

[1]There are a variety of other statistical procedures that could be adopted. For example, in 1998, CARB completed a similar analysis using two methods (CARB 1998). One involved a comparison of arithmetic-averages without estimating uncertainty. The other was a more-rigorous statistical approach that analyzed effects due to differences in vehicles as well as effects due to differences in fuel composition. Both approaches yielded conclusions that are very similar to the ones presented here.

TABLE 7-1 Properties of the Fuels Selected for the Case Study[a]

Fuel	Ethanol (vol%)	MTBE (vol%)	Oxygen (wt%)	RVP (psi)[b]
AQIRP Phase I[c]				
F	0	0	0	8.8
S	0	0	0	8.0
U	9.7	0	3.4	9.6
T	9.7	0	3.4	9.3
N2	0	14.5	2.6	8.8
MM	0	14.8	2.7	8.0
AQIRP Phase II[d]				
C1	0	0	0	6.9
C2	0	11.2	2.0	6.8
California Ethanol Testing Program[e]				
63	0	11.6	2.1	6.9
64	11.2	0	3.9	7.8

[a]See Table 6-1 for a more-detailed tabulation of the fuel properties.
[b] RVP (psi), Reid vapor pressure (pounds per square inch).
[c]Fuel benzene, 1.4 ± 0.1 vol%; aromatics, 19.1-22.2 vol%; alkenes, 3.1-5.4 vol%; sulfur, 246-345 parts per million (ppm by wt).
[d]Fuel benzene, 0.93-0.94 vol%; aromatics, 22.7-25.4 vol%; alkenes, 4.1-4.6 vol%; sulfur, 31-38 ppm by wt.
[e]Fuel benzene, 0.82-0.83 vol%; aromatics, 23.3 vol%; alkenes, 4.8-4.9 vol%; sulfur, 32-34 ppm by wt.

the available database did not allow such a straightforward comparison; the fuels chosen were the best available to the committee and differ in the percent (by weight) of oxygen and the percent (by volume) of ethanol compared with MTBE (see Table 7-1).

ASSESSING WHETHER EMISSIONS AND REACTIVITY DIFFERENCES ARE STATISTICALLY SIGNIFICANT

As discussed in Chapter 3, the calculation of reactivity for any given VOC or combination of VOCs can be in error for any number of reasons (e.g., errors in the chemical mechanism used to calculate the reactivity factors, or errors in the speciation assumed for the VOC mixture). As a result, there is an uncertainty associated with the reactivity calculated for the emissions from any source, including those arising from the individual LDV using various blends of RFGs. The magnitude of the uncertainty in these reactivities is a crucial piece of information needed to decide whether one RFG blend is preferable over another from an air-quality point of view.

The uncertainty in any measured parameter, including those related to LDV emissions, can arise from both random and systematic errors. Systematic error is defined as the difference between the true value of the quantity of interest and the value to which the mean of the measurements converges as more measurements are taken. These types of errors can arise from faulty experimental protocols or incorrect model assumptions, and introduce a bias into the results. Scientists and engineers always seek to eliminate all systematic errors. Nevertheless, the possibility of unidentified systematic errors can rarely be totally eliminated and, because they are often unidentified, they are difficult to quantify.

Random errors are somewhat easier to characterize by adopting a probabilistic or statistical approach. For example, take fuel *a* and fuel *b* and suppose that each fuel is tested on *m* different vehicles. On the basis of these *m* tests, the mean (or average) reactivity for each fuel can be calculated from

$$\overline{R_x} = \frac{1}{m} \sum_{i=1}^{m} (R_x)_i \, ,$$

(7-1)

where $\overline{R_x}$ is the mean reactivity for fuel *x* (*x* = *a* or *b*), and $(R_x)_i$ is the reactivity for fuel *x* obtained from test *i*. The variance is estimated by

$$(s_x)^2 = \frac{1}{(m-1)} \sum_{i=1}^{m} [(R_x)_i - \overline{R_x}]^2$$

(7-2)

Together $\overline{R_x}$ and $(s_x)^2$ describe the probability that a new measurement of R_x will have a specific value, with $\overline{R_x}$ being the most probable value and s_x describing the spread of values about $\overline{R_x}$. When the probability can be described by a probability density function (as in Figure 7-1),

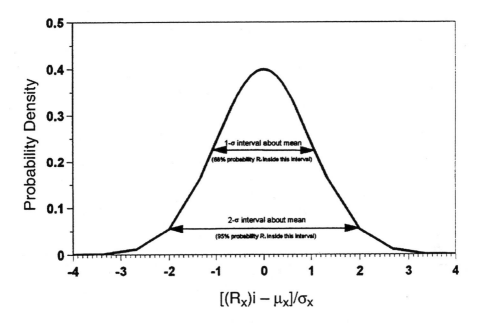

FIGURE 7-1 The probability density distribution for a population of reactivities, R_x for fuel X about the population mean (μ_x) with a variance given by $(\sigma_x)^2$.

there is a 68% probability that an additional measurement of R_x will lie between $\overline{R_x} - s_x$ and $\overline{R_x} + s_x$ and a 95% probability the measurement will lie between $\overline{R_x} - 2s_x$ and $\overline{R_x} + 2s_x$.

Although s_x defines the spread in the population of measured values of R_x, it does not define the uncertainty with which the mean reactivity, $\overline{R_x}$, is defined. To do this, the standard deviation of the mean[2] is used:

$$s_{m,x} = \frac{s_x}{\sqrt{m}}. \qquad\qquad (7\text{-}3)$$

---

[2]In this report, "uncertainty" and the "standard deviation of the mean" are used interchangeably. It should be borne in mind, however, that this metric of uncertainty only includes that arising from random errors and not those from systematic errors.

The 1-s confidence interval (i.e., the interval between $\overline{R_x} - s_{m,x}$ and $\overline{R_x} + s_{m,x}$) will contain the true or actual reactivity of fuel x 68% of the time and the $2s_{m,x}$ confidence interval will contain the true reactivity 95% of the time.

For a decision maker confronted with choosing between two RFG blends on the basis of their reactivity, a critical question is whether the difference in the two reactivities is statistically significant. The answer to this question is closely tied to the magnitude of the standard deviation of the mean, $s_{m,x}$, for the two fuels. The smaller the values for $s_{m,x}$, the greater the likelihood of being able to establish that a small difference in reactivities is statistically significant. Thus, by inspection of Equation 7-3, we see that the most useful emissions studies for this purpose are those that involve a large number of (vehicle) tests and minimize the sources of random experimental error (e.g., from temporal fluctuations in laboratory conditions).

However, simply knowing the magnitudes of the $s_{m,x}$ values does not in and of itself provide the answer to the question of statistical significance. A set of rules must be adopted for deciding whether any similarity or difference in the reactivities of two RFG blends is in fact statistically significant. Typically, these rules include an appropriate type of statistical test and a selection of the level of confidence that will be required to certify statistical significance. Although the statistical test is an objective procedure, the setting of the level of confidence is a more-subjective exercise that relates to the concerns and priorities of the decision maker. In general, the decision maker must decide whether it is more important to avoid falsely concluding that a difference exists or to avoid falsely concluding that no difference exists. If a decision maker uses a difference in the mean reactivities measured for two fuels to implement a given control policy (e.g., choosing fuel a over fuel b on the basis of experimental data) but, in fact, there is no difference in the real world, the decision maker has committed a Type I error (falsely concluding that a difference exists). Such an error might not have a negative impact on air quality, but it could very well incur unnecessary economic costs. If on the other hand, the decision maker decides that the two reactivities are not significantly different and thus does not choose fuel a over fuel b when in fact the true reactivities are different, the decision maker has made a Type II error (falsely concluding that no difference exists). In this case, the error could have an unintended negative air-quality impact. Choosing which error is more important to avoid and setting an acceptable level of risk of committing either error are policy decisions. The following discussion illustrates how one statistical approach can incorporate such choices when assessing fuels based on ozone-forming potential.

Consider an experiment in which motor vehicles are randomly selected for emissions testing. Each vehicle will be used to combust fuel a and fuel b, and the reactivities of the emissions are obtained, R_a and R_b respectively. The null hypothesis (typically denoted by statisticians as "H_o") is that there is no difference in the reactivities of emissions from a sampling of vehicles using fuel a versus fuel b, that is, $\mu_a = \mu_b$. The alternative hypothesis (denoted by statisticians as "H_a"), specifies that $\mu_a \neq \mu_b$.

The two-tailed paired t test provides a methodology for determining the confidence or statistical probability that H_o can be rejected in favor of H_a or vice versa. One of the parameters calculated in a paired t test is the so-called "p value." This parameter can vary between 0 and 1 and increases as the difference in the emissions between two fuels becomes smaller and/or less statistically significant. It is defined as the probability that the null hypothesis, H_o is true, and it thus equal to (1 - probability) that H_o is false. Representative p values and the various probabilities implied by these values are listed in Table 7-2. For example, if the p value for a given paired t test is 0.05, there is a 5% probability that the null hypothesis is correct and a 95% probability that the null hypothesis is incorrect. (Another way of stating this is to say that the two reactivities are statistically different at the 95% confidence level.) On the other hand, if the p value for a given test is 0.95, there is 95% probability that the null hypothesis is correct, and so forth.

Because the p value is the probability that the null hypothesis is true, it is equivalent to the probability of making a Type I error (i.e., incorrectly choosing one fuel over another when there is in fact no difference in their emissions). Thus, when a small p value (reflecting large and significant differences in the reactivities of two fuels) is obtained, there is a relatively small probability of making a Type I error. In this case, the decision maker could choose the lower reactivity fuel with a high degree of confidence. On the other hand, when a large p value is obtained, a decision maker is likely to make a Type I error by choosing the fuel with the apparent, but not statistically significant lower reactivity.

In general, as the probability of making a Type I error increases, the probability of making a Type II error (i.e., not choosing the lower reac-

Table 7-2   Representative p Values and Associated Probabilities for a Two-Tailed t Test on Reactivities R_a and R_b

p Value	Probability (%) That $R_a \neq R_b$	Probability (%) of Making Type I Error by Choosing One Fuel over the Other	Probability (%) of Making Type I Error by Not Choosing One Fuel over the Other
0.01	99	1	High
0.05	95	5	
0.1	90	10	
0.15	85	15	
0.2	80	20	
0.4	60	40	Moderate
0.6	40	60	
0.8	20	80	Low

tivity fuel) decreases. Thus, low p values imply a high probability of a Type II error if a decision maker decides to not choose the lower reactivity fuel, while high p values imply a low probability of a Type II error.

FUELS AND EMISSIONS DATA FROM THE AQIRP STUDY

As indicated in Table 7-1, eight fuels from the AQIRP study were selected for detailed analysis here: six from AQIRP Phase I and two from AQIRP Phase II. Collectively, the eight fuels provide a range of properties related to oxygen content and type of oxygenate. Fuel F, used in Phase I of the AQIRP, was an RFG with low aromatic content, low alkene content, low $T_{90,}$ and no oxygen. Fuel S was similar to fuel F, but with less butane, which resulted in a lower Reid vapor pressure (RVP). Approximately 10% ethanol was splash-blended into fuels F and S to form fuels U and T, respectively. As a result of this splash blending, the RVPs for fuels U and T were about 1 pound per square inch (psi) higher than the RVPs of fuels F and S. Fuels N2 and MM, on the other hand, contained oxygen but in the form of MTBE instead of ethanol. The MTBE was fully blended to the specifications of fuels F and S, respectively. As a result, no dilution effect on aromatic content, alkene content, or T_{90} was produced and the RVPs of fuels N2 and MM were identical to those of fuels F and S, respectively (Table 7-1). Fuel C2, used in AQIRP Phase II, was a low-sulfur RFG that contained MTBE and met the

1996 California Phase 2 regulatory requirements (see Chapter 5), whereas fuel C1 was a fuel blended to essentially the same requirements, but without MTBE.

Emissions for all Phase I fuels (F, S, U, T, N2, and MM) were measured using current-fleet vehicles. The Phase II fuels (C1 and C2), on the other hand, were tested using current-fleet vehicles, federal Tier I vehicles, and advanced-technology vehicles (see Chapter 4). It also should be noted that all of the vehicles in the AQIRP study were well-maintained and properly functioning and thus the data do not address the probable substantial contributions from high-emitting vehicles to overall precursor emissions.

Tables listing the LDV emissions from each of these fuels derived from the AQIRP data are presented in Appendix D. These data were gathered using the Federal Test Procedure (FTP) for exhaust and evaporative emissions according to the procedures described by Rueter et al. (1992) for the Phase I fuels and Burns et al. (1995) for the Phase II fuels. In the case of the Phase I fuels, data for exhaust, diurnal, and hot-soak emissions are presented. Although running-loss emissions were also measured for the Phase I study fuels, only a small number of tests were carried out ($n = 6$ for fuel F; 9 for fuel U; and 2 each for fuels S, T, N2 and MM) and the observed variations were very large (e.g., running-loss mass VOC emissions for the six vehicles tested with fuel U varied by a factor of ~2,000). Accordingly, it is unlikely that these data are representative of the on-road vehicle fleet, and thus the running-loss data for these fuels are not considered here. Nevertheless, it should be borne in mind that high running losses due to fuel leakages and improper vehicular maintenance can be an important or even dominant source of VOC emissions from modern vehicles. In the case of the Phase II fuels, diurnal and running-loss emissions were not measured. Moreover, hot-soak emissions from fuel C1 were measured on only one advanced-technology vehicle and only three advanced-technology vehicles for fuel C2. Given this small sample size, the results of the hot-soak-emissions tests for this class of vehicles are not discussed here.

In addition to the mass of emissions, the tables in Appendix D indicate the total and specific reactivities[3] of the emissions. The exhaust-

---

[3]All reactivities discussed here are based on the maximum incremental reactivity (or MIR) scale and are derived using reactivity factors calculated from the SAPRC 1997 chemical mechanism. Similar conclusions are obtained using the SAPRC 1990 and SAPRC 1993 chemical mechanisms (see Chapter 3).

emissions reactivities include carbon monoxide (CO). The committee has found that CO typically contributes 15% to 25% of the total exhaust-emissions reactivity independent of the fleet type (i.e., current, federal Tier 1, or advanced technology). Thus, the contribution of CO to the exhaust reactivity is quite substantial and should not be neglected.

Before turning to an analysis of the differences in the emissions from the various fuels, it is relevant to note the rather large variability in the data from the AQIRP study. Inspection of the tables in Appendix D reveals that the mass of emissions (in units of grams per mile) measured for a given fuel often varied from one vehicle test to another by a factor of two or more and sometimes by more than a factor of five. This variability is perhaps not surprising in light of earlier discussions in this report of the myriad factors that can influence LDV emissions. Nevertheless, this large variability—compounded with the relatively small number of independent tests carried out for each fuel (typically less than 10)—tended to produce relatively large variances in the mean emissions for each fuel.

Given the substantial variability in emissions of the various vehicles tested with fuels A and B, the committee used logarithm[(emissions using fuel A) ÷ (emissions using fuel B)] for each vehicle used in the paired t test. The use of such an approach assumes, reasonably, that substituting fuel A for fuel B causes a constant fractional (or percentage) change in the emissions being considered (CO, NO_x, VOC, etc.). When a number of tests was available for a given vehicle–fuel combination, an arithmetic mean was used for input into the logarithm [(emissions using fuel A) ÷ (emissions using fuel B)]. Obviously, only vehicles for which emissions tests were carried out using both fuels could be used in the paired t-test statistical analysis.

Effect of Reid Vapor Pressure

In addition to affecting the oxygen content, the presence of oxygenates (and especially ethanol) in gasoline can increase the fuel's RVP. Moreover, a primary effect of increasing the RVP of gasoline is to increase the evaporative emissions from LDVs. In the committee's assessment of the impact of oxygenates on RFG emissions, it would be useful, therefore, if one could separate out the effect of RVP increases from that of the addition of oxygen. Toward that end, it is instructive to assess what effect increased RVP in the AQIRP fuels had on the emissions measured during that study. Inspection of Table 7-1 indicates that there are three fuel pairs with very similar properties except for their RVPs; comparison of the emissions from these pairs thus provides an opportunity to assess the effect of RVP observed by AQIRP. The fuel pairs are

    1.   Fuel S (oxygen = 0%, RVP = 8.0 psi) vs. fuel F (oxygen = 0%, RVP = 8.8 psi).
    2.   Fuel MM (oxygen = 2.7% using MTBE, RVP = 8.0 psi) vs. fuel N2 (oxygen = 2.6% using MTBE, RVP = 8.8 psi).
    3.   Fuel T (oxygen = 3.4% using ethanol, RVP = 9.3 psi) vs. fuel U (oxygen = 3.4% using ethanol, RVP = 9.6 psi).

The reader will note that while the first two fuel pairs have a 0.8-psi difference in RVP, the third pair has only a 0.3-psi difference in RVP. Thus, if RVP has an effect on emissions, one might expect to find a larger difference in the emissions from the first two pairs compared with the third.

    A comparison of the exhaust, diurnal, and hot-soak emissions of these fuel pairs, and the statistical significance of the differences in terms of the p values are presented in Tables 7-3, 7-4, and 7-5, respectively. Little evidence of a statistically significant effect of RVP is seen from these tables. In most cases, the p values were well above the 0.05 threshold to establish 95% confidence. However, Table 7-3 indicates a consistent decrease in CO emissions for the use of lower-RVP fuels. This observation is in agreement with the findings of Reuter et al. (1992), whose findings were based on eleven fuels in the AQIRP Phase I study (including the six used here) selected to investigate the effects of RVP and oxygenates on vehicular emissions. Reuter et al. (1992) found that, after combining the results from non-oxygenated fuels with fuels containing MTBE or ethanol, a 1.0-psi reduction in RVP resulted in a reduction in exhaust CO emissions of 9.1% (significant at the 95% confidence levels). On the other hand, some unexpected (even curious) results appear. For example, although the major effect of lower RVP is thought to be to lower evaporative emissions, the data presented here by no means confirm this trend. In fact, for each emissions category, lower RVP is associated with higher diurnal or hot-soak emissions in at least one of the three fuel pairs considered here. In the case of hot-soak emissions, a lower RVP fuel produced a higher reactivity that was significant at the 93% confidence level.

TABLE 7-3   Effect of RVP on Exhaust Emissions from Three AQIRP Fuel Pairs

Fuel Pair	% Decrease in Emissions Attributable to Lower RVP[1]	p Value[2]	Summary
A. Effect on mass of VOC emissions (g/mi)			
S/F	-9	0.2	Data from fuel pairs inconsistent. In first case, ~80% probability that higher RVP fuel has lower emissions; in the second, >95% probability that lower RVP has lower emissions; and in the third, >60% probability that RVP has no effect on emissions.
MM/N2	11	0.04	
T/U	1	0.6	
B. Effect on mass of CO emissions (g/mi)			
S/F	7	0.3	Data indicate >40% probability that lower RVP fuel has lower emissions. Probability of Type I error small.
MM/N2	19	0.009	
T/U	10	0.4	
C. Effect on total reactivity (g O_3/mi)			
S/F	-7	0.5	No consistent, statistically significant effect apparent.
MM/N2	12	0.2	
T/U	1	0.70	
D. Effect on mass of NO_x emissions (g/mi)			
S/F	-7	0.1	No significant and consistent effect apparent.
MM/N2	6	0.8	
T/U	~0	0.2	

[1]% decrease = [(emissions with low RVP) − (emissions with high RVP)] ÷ (emissions with high RVP). Negative value indicates an emissions or reactivity increase with lower RVP.

[2]Based on logarithms of means.

Effect of Oxygenates Using MTBE

Inspection of Table 7-1 indicates that there are three fuel pairs that can be used to assess the effect of adding MTBE to gasoline, two from AQIRP Phase I and one from AQIRP Phase II. In each case, the fuel pairs have essentially identical RVPs, and as a result, the comparisons are well-

TABLE 7-4 Effect of RVP on Diurnal Emissions from Three AQIRP Fuel Pairs

Fuel Pair	% Decrease in Emissions Attributable to Lower RVP[1]	p Value[2]	Summary
A. Effect on mass of VOC emissions (g/test)			
S/F	36	0.02	Nonoxygenated fuels show significant decrease in emissions with lower RVP. However, data for oxygenated fuels do not consistently confirm this trend.
MM/N2	-3	0.3	
T/U	16	0.2	
B. Effect on total reactivity (g O_3/test)			
S/F	29	0.01	Nonoxygenated fuels show significant decrease in reactivity of emissions with lower RVP. However, data for oxygenated fuels do not consistently confirm this trend.
MM/N2	-8	0.1	
T/U	13	0.3	

[1]% decrease = [(emissions with low RVP) – (emissions with high RVP)] ÷ (emissions with high RVP). Negative value indicates an emissions or reactivity increase with lower RVP.
[2]Based on logarithms of means.

TABLE 7-5 Effect of RVP on Hot-Soak Emissions from Three AQIRP Fuel Pairs

Fuel Pair	% Decrease in Emissions Attributable to Lower RVP[1]	p Value[2]	Summary
A. Effect on mass of VOC emissions (g/test)			
S/F	16	0.7	No consistent statistically significant effect of RVP on emissions is apparent.
MM/N2	-20	0.1	
T/U	9	0.03	
B. Effect on total reactivity (g O_3/test)			
S/F	16	0.8	No consistent statistically significant effect of RVP on reactivity of emissions is apparent.
MM/N2	-31	0.07	
T/U	7	0.1	

[1]% decrease = [(emissions with low RVP) – (emissions with high RVP)] ÷ (emissions with high RVP). Negative value indicates an emissions or reactivity increase with lower RVP.
[2]Based on logarithms of means.

suited to isolating the effect of the oxygenated additive. The fuel pairs are

    1.   Fuel N2 (oxygen = 2.6% using MTBE, RVP = 8.8 psi) vs. fuel F (oxygen = 0%, RVP = 8.8 psi).
    2.   Fuel MM (oxygen = 2.7% using MTBE, RVP = 8.0 psi) vs. fuel S (oxygen = 0%, RVP = 8.0 psi).
    3.   Fuel C2 (oxygen = 2 % using MTBE, RVP = 6.8 psi) vs. fuel C1 (oxygen = 0 %, RVP = 6.9 psi).

With regard to fuel pair C2 and C1, it should be noted that (1) comparisons can be made with three types of vehicles—current fleet, Tier I, and advanced technology—and (2) no data are available for diurnal emissions.
    A comparison of the exhaust, diurnal, and hot-soak emissions of these fuel pairs, and the statistical significance of the differences are presented in Tables 7-6, 7-7, and 7-8, respectively. As in the previous comparisons, there is little evidence here to suggest a statistically significant effect of MTBE. In most of the emissions categories, the various fuel pairs produced conflicting results, with the MTBE fuel having lower emissions (or reactivity) in some cases and higher emissions (or reactivity) in other cases. The most consistent effect is an increase in NO_x exhaust emissions from MTBE (Table 7-6D). There are also suggestions of an increase in the mass and reactivity of hot-soak VOC emissions (Table 7-8B), as well as a decrease in CO exhaust emissions (Table 7-6B) from the addition of MTBE.

Effect of Ethanol vs. MTBE

There are two fuel pairs (both from AQIRP Phase I) that provide an indication of the relative effects of using MTBE or ethanol as an oxygenate in RFG. These are

    1.   Fuel T (oxygen = 3.4% using ethanol, RVP = 9.3 psi) vs. fuel MM (oxygen = 2.7% using MTBE, RVP = 8.0 psi).
    2.   Fuel U (oxygen = 3.4% using ethanol, RVP = 9.6 psi) vs. fuel N2 (oxygen = 2.6% using MTBE, RVP = 8.8 psi).

Note that in both cases, the ethanol-blended fuel had about the same

TABLE 7-6 Effect of MTBE on Exhaust Emissions from Three AQIRP Fuel Pairs

Fuel Pair	% Decrease in Emissions Attributable to MTBE[1]	p Value[2]	Summary
A. Effect on mass of VOC emissions (g/mi)			
N2/F	-2	0.9	Fuel-pair MM/S indicates a signifi-
MM/S	16	0.005	cant benefit of adding MTBE. How-
C2/C1			ever, no statistically significant ef-
Current fleet	-1	0.5	fect is apparent from other fuel
Tier I	2	0.6	pairs.
Adv. technol.	6	0.3	
B. Effect on mass of CO emissions (g/mi)			
N2/F	-2	0.2	No consistent, statistically signifi-
MM/S	11	0.4	cant effect is apparent.
C2/C1			
Current fleet	10	0.7	
Tier I	1	0.2	
Adv. technol.	7	0.3	
C. Effect on total reactivity (g O_3/mi)			
N2/F	-5	0.8	No consistent, statistically signifi-
MM/S	14	0.07	cant effect is apparent.
C2/C1			
Current fleet	1	0.2	
Tier I	3	0.6	
Adv. technol.	6	0.2	
D. Effect on mass of NO_x emissions (g/mi)			
N2/F	-17	0.05	Data suggest an increase in NO_x
MM/S	-3	0.4	emissions from the addition of
C2/C1			MTBE in all but advanced-technol-
Current fleet	-6	0.7	ogy vehicles. Likelihood of Type I
Tier I	-11	0.1	error is small.
Adv. technol.	2	0.9	

[1]% decrease = [(emissions with MTBE) − (emissions without MTBE)] ÷ (emissions without MTBE). Negative value indicates an emissions or reactivity increase with the addition of MTBE.

[2]Based on logarithms of means.

TABLE 7-7 Effect of MTBE on Diurnal Emissions from Three AQIRP Fuel Pairs

Fuel Pair	% Decrease in Emissions Attributable to MTBE[1]	p Value[2]	Summary
A. Effect on mass of VOC emissions (g/test)			
N2/F	29	0.03	Data are not consistent.
MM/S	-15	0.3	
C2/C1	No data	–	
B. Effect on total reactivity (g O_3/test)			
N2/F	13	0.1	Data do not indicate a consistent
MM/S	-33	0.06	effect.
C2/C1	No data	–	

[1]% decrease = [(emissions with MTBE) – (emissions without MTBE)] ÷ (reactivity without MTBE). Negative value indicates an emissions or reactivity increase with addition of MTBE.

[2]Based on logarithms of means.

oxygen content as that of the MTBE-blended fuel and thus less volume percent oxygenate (see Table 5-2). A comparison of ethanol and MTBE-blended fuels with similar volume percent oxygenate, but different oxygen content, is provided by the data from the California Ethanol Test Program discussed in the next section. It is also relevant to note that the ethanol-blended fuels had about a 1-psi higher RVP than the MTBE-blended fuel.

A comparison of the exhaust, diurnal, and hot-soak emissions of these fuel pairs, and the statistical significance of the differences are presented in Tables 7-9, 7-10, and 7-11, respectively. As in the previous two comparisons, the data presented here from the AQIRP study on the relative benefits of ethanol and MTBE are by no means conspicuous or striking. There is a suggestion that ethanol (or the higher RVP it engendered in the fuels considered here) caused somewhat higher VOC exhaust and evaporative emissions. In each case, however, the effect of ethanol on the reactivity of the emissions was less than its effect on the mass of the VOC emissions. Finally it is relevant to note that an analysis of air toxic emissions from the AQIRP fuels considered here suggests that there are advantages and disadvantages related to the use of either oxygenate (see Text Box 7-1). The above conclusions concerning the

TABLE 7-8   Effect of MTBE on Hot-Soak Emissions from Three AQIRP Fuel Pairs

Fuel Pair	% Decrease in Emissions Attributable to MTBE[1]	p Value[2]	Summary
A. Effect on mass of VOC emissions (g/test)			
N2/F	8	0.10	Data are not consistent but, over-
MM/S	−32	0.02	all, indicate a small probability that
C2/C1			addition of MTBE causes an in-
Current fleet	−23	0.4	crease in hot-soak emissions.
Tier I	−17	0.8	
Adv. technol.	Insufficient data	−	
B. Effect on total reactivity (g O₃/test)			
N2/F	12	0.7	Data are not consistent but, over-
MM/S	−38	0.02	all, indicate a small probability that
C2/C1			addition of MTBE causes an in-
Current fleet	−11	0.9	crease in the reactivity of hot-soak
Tier I	−8	0.5	emissions.
Adv. technol.	Insufficient data	−	

[1]% decrease = [(emissions with MTBE) − (emissions without MTBE)] ÷ (emissions without MTBE). Negative value indicates an emissions or reactivity increase with addition of MTBE.
[2]Based on logarithms of means.

effects of RVP, MTBE, and ethanol on vehicle emissions are generally consistent with those reported by Dunker et al. (1996) from a modeling study of the impacts of different gasoline fuels on ozone levels in the Los Angeles, Dallas–Ft. Worth, and New York urban areas using the AQIRP data and the Urban Airshed Model.

FUELS AND EMISSIONS DATA FROM THE CALIFORNIA ETHANOL TESTING PROGRAM

Because of the limited number of tests made in AQIRP that directly compared emissions from MTBE-containing and ethanol-containing fuels, the data from that study provide only fragmentary information on the

TABLE 7-9   Effect of Ethanol vs. MTBE on Exhaust Emissions from Two AQIRP Fuel Pairs

Fuel Pair	% Decrease in Emissions Attributable to Ethanol[1]	p Value[2]	Summary
A. Effect on mass of VOC emissions (g/mi)			
T/MM	-11	0.04	Some indication that ethanol might
U/N2	~0	0.08	cause higher VOC mass emissions.
B. Effect on mass of CO emissions (g/mi)			
T/MM	-11	0.6	No consistent, statistically signifi-
U/N2	1	0.4	cant effect is apparent.
C. Effect on total reactivity (g O_3/mi)			
T/MM	-8	0.4	No consistent, statistically signifi-
U/N2	5	0.4	cant effect is apparent.
D. Effect on mass of NO_x emissions (g/mi)			
T/MM	2	0.4	No statistically significant effect is
U/N2	7	0.4	apparent.

[1]% decrease = [(emissions with ethanol) − (emissions with MTBE)] ÷ (emissions with MTBE). Negative value indicates an emissions or reactivity increase with the addition of ethanol.
[2]Based on logarithms of means.

TABLE 7-10 Effect of Ethanol vs. MTBE on Diurnal Emissions from Two AQIRP Fuel Pairs

Fuel Pair	% Decrease in Emissions Attributable to Ethanol[1]	p Value[2]	Summary
A. Effect on mass of VOC emissions (g/mi)			
T/MM	-12	0.43	Data indicate probability that etha-
U/N2	-38	0.01	nol causes higher mass emissions.
B. Effect on total reactivity (g O_3/mi)			
T/MM	5	0.81	Data are not consistent, but most
U/N2	-18	0.02	likely effect is an increase in reactiv-ity of emissions from ethanol.

[1]% decrease = [(emissions with ethanol) − (mission with MTBE)] ÷ (emissions with MTBE). Negative value indicates an emissions or reactivity increase with the addition of ethanol.
[2]Based on logarithms of means.

TABLE 7-11 Effect of Ethanol vs. MTBE on Hot-Soak Emissions from Two AQIRP Fuel Pairs

Fuel Pair	% Decrease in Emissions Attributable to Ethanol[1]	p Value[2]	Summary
A. Effect on mass of VOC emissions (g/mi)			
T/MM	−14	0.29	Data indicate > 30% probability that ethanol causes higher mass emissions.
U/N2	−50	0.003	
B. Effect on total reactivity (g O_3/mi)			
T/MM	1	0.72	Data are not consistent, but the most likely effect is an increase in reactivity of emissions from ethanol.
U/N2	−40	0.002	

[1]% decrease = [(emissions with ethanol) − (emissions with MTBE)] ÷ (emissions with MTBE). Negative value indicates an emissions or reactivity increase with the addition of ethanol.

[2]Based on logarithms of means.

TEXT BOX 7-1   Effect of Oxygenates on Toxic Air Contaminant Emissions

Exhaust and evaporative emissions of selected air toxics from LDVs using the six AQIRP Phase I fuels are listed in Table 7-12. The data suggest that the fuels result in similar emissions of 1-3 butadiene and benzene (i.e., they fall within the observed variability as indicated by the standard deviations of the means). However, there appear to be differences in acetaldehyde and formaldehyde emissions that at least border on being larger than the observed variability. In the case of acetaldehyde exhaust emissions, the ethanol-containing fuels produce about a factor of 2 larger exhaust emissions than that of the MTBE-containing and oxygen-free fuels. On the other hand, the ethanol-containing fuels tend to result in somewhat lower exhaust emissions of formaldehyde. It is also interesting to note that while MTBE-containing fuels are generally thought to result in enhanced exhaust emissions of formaldehyde (see Chapter 6), this trend is not reflected in the data presented in Table 7-12.

TABLE 7-12 Toxic Air Contaminant Emissions from AQIRP Phase I Fuels

Fuel	Exhaust Formaldehyde (mg/mi)	Exhaust Acetaldehyde (mg/mi)	Exhaust 1,3- Butadiene (mg/mi)	Exhaust Benzene (mg/mi)	Diurnal Benzene (mg/test)	Hot-Soak Benzene (mg/test)
F (0% oxygen, RVP = 8.8 psi)	1.65 ± 0.6	0.88 ± 0.3	0.70 ± 0.2	8.1 ± 3	7.8 ± 4	15.0 ± 4
S (0% oxygen, RVP = 8.0 psi)	1.40 ± 0.6	0.88 ± 0.6	0.82 ± 0.3	8.8 ± 5	6.2 ± 4	11.2 ± 5
U (3.4% oxygen using EtOH, RVP = 9.6 psi)	0.99 ± 0.3	1.38 ± 0.4	0.73 ± 0.3	7.7 ± 4	7.8 ± 6	16.2 ± 6
T (3.4% oxygen using EtOH, RVP = 9.3 psi)	1.13 ± 0.4	2.07 ± 1.0	0.76 ± 0.3	7.8 ± 4	7.0 ± 6	15.6 ± 6
N2 (2.6% oxygen using MTBE, RVP = 8.8 psi)	1.28 ± 0.6	0.74 ± 0.5	0.81 ± 0.4	8.3 ± 4	5.5 ± 6	10.1 ± 6
MM (2.7% oxygen using MTBE, RVP = 8.0 psi)	1.71 ± 0.9	0.84 ± 0.6	0.73 ± 0.3	6.7 ± 3	5.7 ± 4	12.1 ± 5

Uncertainties are 1 standard deviation of the mean.

relative benefits of the two types of oxygenates in RFG. Fortunately, the California Ethanol Testing Program produced a considerably more-detailed database on this issue. During the program, multiple tests were made to characterize the emissions from LDVs using a fuel with MTBE (fuel 63) and a comparable fuel with ethanol (fuel 64). As indicated in Table 7-1, fuel 63 contained 2.1% oxygen by weight from MTBE and had an RVP of 6.9 psi; fuel 64 contained 3.9% oxygen by weight from ethanol and had an RVP of 7.8 psi. In all other respects (e.g., benzene, aromatic, and sulfur content), the fuels were essentially identical. Thus, a comparison of the two fuels directly addresses the question of whether the tendency for ethanol to increase RVP can be overcome by the addition of more oxygen.

Exhaust emissions in the program were measured using the FTP (Calvert et al. 1993) and the Rep05 test procedure.[4] The evaporative emissions measured were hot-soak and 0-24-hr and 24-48-hr diurnal emissions (see Chapter 4). No running-loss emissions were measured but were estimated using an emissions model. Because this model is proprietary, its performance could not be assessed by this committee, and as a result, these emissions estimates are not included in this report. In addition to measurements of the mass of VOC, NO_x, CO, and toxic emissions, the emissions of hydrocarbons, alcohols, carbonyls, and aldehydes were speciated; thus making possible reactivity calculations for the nonmethane organic gases (NMOGs). A detailed summary of the reactivities of the exhaust and evaporative emissions from fuels 63 and 64 are presented in Appendix D. These reactivities were calculated using the MIR scale. The reactivity factors used are found in the California Test Procedure, adopted by CARB in July 1992 and last amended June 24, 1996.

Fourteen vehicles of model years 1990 to 1995 were selected and classified according to engine family, evaporative family, and emissions-control technologies and then used to characterize emissions from the fuels. Acceptance criteria for each vehicle were based upon a protocol developed by the In-Use Compliance Section of CARB. This protocol consisted of a telephone questionnaire, a 10-point inspection of the vehicle, and restorative maintenance. The purpose of restorative maintenance was to bring the vehicle into manufacturer's specification and to ensure that all electrical and mechanical controls are functioning prop-

---

[4]The REP05 is a high-speed, high-acceleration test procedure (CARB 1998).

erly. Further requirements were that the vehicles pass a smog check, not exceed specified mileage limits set for the different model years included in the study, and it be obtained from rental fleets rather than private owners if possible.

Exhaust and evaporative emissions from each of the 14 vehicles with both fuels 63 and 64 were measured two times, and in some cases, three times. (However, tests involving two of the vehicles were discarded due to nonstatistical errors.) Thus, the mean emissions from each of the fuels for each emissions category were derived from almost 30 separate tests, a much larger number than that typically used to derive the mean emissions from the AQIRP data discussed in the previous sections. The larger number of tests in the California Ethanol Testing Program should make these data more amenable to discerning subtle differences in the emissions from each fuel.

The analyses suggest that the reactivity of the exhaust emissions for the ethanol-blended fuel was about 4% less than that of the MTBE-blended fuel. That decrease is essentially all attributable to an approximate 10% decrease in the mass of CO exhaust emissions for the ethanol-blended fuel. However, this relatively small decrease in the reactivity of the exhaust emissions was overwhelmed by the much larger increase in the mass and reactivity of the evaporative VOC emissions arising from the use of ethanol-blended fuel. As a result, the reactivity of the combined exhaust and evaporative emissions using the ethanol-blended fuel was estimated by CARB to be about 17% larger than those using the MTBE-blended RFG. The committee analyzed data obtained from the California Ethanol Testing Program before publication of CARB's (1998) analysis and before data on the reactivity of CO emissions were available. The committee compared the reactivities of emissions from fuels 63 and 64 using a two-sample *t* test (see Table 7-13 and Appendix D). Since the committee completed its analysis, CARB published its results of a more-comprehensive analysis of the data from the California Ethanol Testing Program. Although the results of CARB's analysis are somewhat different from those of the committee, the overall conclusions are the same.

IS THERE A DIFFERENCE BETWEEN CONCLUSIONS DRAWN ON THE BASIS OF VOC-MASS EMISSIONS AND THE REACTIVITY OF THE EMISSIONS?

In Chapter 3, we noted that because of the wide range of VOC species typically emitted by LDVs and the highly variable chemistry of these

TABLE 7-13 Effect of Ethanol vs. MTBE on Total Reactivity (g O_3/mi or g O_3/test) of Emissions from Fuels 63 and 64 of the California Ethanol Testing Program

Emissions Type	% Decrease in Reactivity Attributable to Ethanol[1]	p Value[2]	Statistical Summary
Exhaust based on FTP composite	-9	0.50	No consistent, statistically significant effect is apparent.
Exhaust based on Rep-05	5	0.3	
Hot Soak	-73	0.002	>99% probability that difference in reactivity is significant.
0-24 Diurnal	-60	0.004	>99% probability that difference in reactivity is significant.
24-48 Diurnal	-82	0.002	>99% probability that difference in reactivity is significant.

[1]% decrease = [(emissions with ethanol) − (emissions with MTBE)] ÷ (emissions with MTBE). Negative value indicates an emissions or reactivity increase with the addition of ethanol.

[2]Exhaust reactivities did not include CO.

compounds, the mass of VOC emissions might be a poor metric for the ozone-forming potential of these emissions. Under some circumstances a reactivity scale might provide a more-reliable measure. In light of this situation, it is interesting to consider whether the conclusions drawn above with regard to the relative benefits of ethanol and MTBE are affected by which metric is used.

Inspection of the data in Tables 7-3 through 7-13, as well as those provided in Appendix D, suggest that the two metrics did in fact produce some differing results. For example, note in Tables 7-10 and 7-11 that the mass of evaporative emissions from AQIRP fuels with ethanol are greater than those from fuels with MTBE. However, for one of the fuel pairs considered, the difference is cut by more than a factor of 2 when measured on the basis of reactivity; in the case of the other fuel pair, the reactivity from the ethanol-containing fuel is actually found to be less than that of the MTBE-containing fuel. However, in this latter case, the difference in both the mass and reactivity of emissions was not statistically significant.

A contrasting result was obtained for hot-soak emissions from the fuels in the California Ethanol Testing Program. In this case, the ethanol-containing fuel also has a larger mass of emissions than the MTBE-containing fuel. However, this difference was further enhanced when the reactivity of the emissions was considered.

Despite these differences, however, it is important to note that, in none of these instances did the use of the reactivity metric fundamentally alter the conclusions that would have been reached if the mass-emissions metric had been adopted. For example, note in Table 7-10 that fuel U (containing ethanol) was found to produce higher emissions than fuel N2 (containing MTBE) at a greater than 95% confidence level regardless of the metric used; the inconsistency between the two metrics is only in the magnitude of the difference between the fuels. In the case of fuels T and MM on the other hand, the mass-emissions metric indicates higher emissions for the ethanol-containing fuel while the reactivity metric indicates lower emissions for the ethanol-containing fuels. However, in these cases the *p* values are relatively large, and thus the differences in the mass and reactivity of the emissions from two fuels are not statistically significant.

ANALYSIS USING THE COMPLEX AND PREDICTIVE MODELS

The analyses presented in the preceding two sections could perhaps be criticized for being based on test results from a limited number of fuels, and thus not representative of a fleet-wide response to changes in fuel composition. Indeed, in the case of the AQIRP study the smallness of the sample size limited the ability to unequivocally conclude that oxygenates had, or did not have, an impact. Other researchers (Mayotte et al. 1994) also find some indications of an impact, but warn that their sample size was limited as well.

As noted in Chapter 4, both EPA and CARB have conducted statistical analyses of a much larger number of tests to develop models to predict how the mass of VOC and NO_x emissions respond to fuel-composition changes. (Recall that EPA's model is called the Complex Model, and CARB's is called the Predictive Model). The databases used to develop both models are similar. The major differences are in the statistical treatment of the data, and that the Complex Model has a separate segment for high-emitting vehicles (CARB 1991).

As a final check on the applicability of the results discussed above, the properties of the 10 fuels listed in Table 7-1, as well as the California Phase 2 reference fuel, were input into the Complex and Predictive

Models. The resulting exhaust and evaporative emissions predicted by the Complex Model are given in Table 7-14 and the percentage decrease in exhaust emissions predicted by the Predictive Model, relative to the reference fuel, are listed in Table 7-15.[5] Because neither the Complex nor the Predictive Models estimate the composition of the emissions, these models cannot be used to predict changes in the reactivity of the emissions.

TABLE 7-14 VOC and NO_x Emissions for Various Fuels Predicted by EPA's Complex Model[a]

Fuel	Emissions (mg/mi)		
	NO_x	Exhaust VOC	Evaporative VOC
C1 (low sulfur)	561	375	370
C2 (low sulfur, MTBE)	563	372	355
MM (MTBE)	639	414	585
N2 (MTBE)	633	422	798
T (ethanol, high RVP)	615	420	956
U (ethanol, high RVP)	625	430	1060
S	628	414	585
F	627	425	798
CA 64 (low sulfur, ethanol)	571	362	539
CA 63 (low sulfur, MTBE)	567	355	365
CA Phase 2 reference	569	367	385

[a]The results in this table are based on the Phase I Complex Model, which contains a higher weighting for evaporative VOC emissions than does the Phase II Complex Model. Therefore, the effects attributable to RVP are expected to be somewhat larger than the effects that would be observed from the Phase II Complex Model. However, the trends among fuels are expected to be similar.

---

[5]The Complex Model calculates the mass of exhaust and evaporative emissions, and the Predictive Model only calculates the percentage decrease in exhaust emissions relative to the reference fuel. The Predictive Model does not consider evaporative emissions.

TABLE 7-15 VOC, CO, and NO_x Exhaust-Emissions Changes for Various Fuels Predicted by CARB's Predictive Model and Draft CO Model[a]

Fuel	% Change from the CA Phase 2 Reference Fuel		
	NO_x	VOC	CO
C1 (low sulfur)	−1.9	2.2	7.4
C2 (low sulfur, MTBE)	−2.0	−3.3	−1.7
MM (MTBE)	17.5	14.3	16.1
N2 (MTBE)	20.7	11.7	19.3
T (ethanol, high RVP)	25.2	5.0	9.6
U (ethanol, high RVP)	28.6	7.7	12.7
S	12.0	19.2	26.2
F	15.4	17.5	28.5
CA 64 (low sulfur ethanol)	8.8	−10.0	−6.1
CA 63 (low sulfur, MTBE)	−1.2	−5.0	−3.0

[a]VOC and NO_x values were provided by K. Cleary of CARB in 1999, using a draft version of the Predictive Model that accounts for RVP changes. CO values are from CARB's draft CO model.

Turning first to the results from the Complex Model, one finds two striking results: (1) the sizable reductions in exhaust emissions arising from low sulfur fuels; and (2) the increase in evaporative emissions with ethanol-containing fuels (presumably from the increased RVP of these fuels). This later result is far more definitive than, although not inconsistent with, the effect of ethanol discerned from the direct analysis of the AQIRP data discussed earlier in this chapter. On the other hand, the small and borderline significant increases in NO_x exhaust emissions and evaporative VOC emissions, as suggested in the AQIRP data, associated with the addition of MTBE are not reflected in the results of the Complex Model.

Like the Complex Model results, the Predictive Model indicates that reducing sulfur content reduces emissions of all components. The model also projects a decrease in CO emissions from the addition of oxygen—an effect that was also seen in the analysis of the emissions data from the AQIRP fuels.

SUMMARY

An analysis of emissions data and regression-model predictions for a limited set of RFG blends with a range of properties, that include different oxygen contents of MTBE and ethanol, suggests that:

- The differences inferred between the VOC emissions of two fuels using the mass of emissions as a metric varied on occasion with that inferred using reactivity as a metric. In some cases, consideration of reactivity decreased the apparent emissions difference and in other cases it enhanced the difference. However, in no case did the fundamental conclusion concerning the choice of one fuel over another (for the fuels studied here), on the basis of statistically significant air-quality benefits, change as a result of using a mass-emissions or reactivity-weighted metric.
- CO emissions account for 15% to 25% of the reactivity of exhaust emissions from LDVs and thus should be included in reactivity assessments because CO contributes to ozone formation due to its large amount of emissions.
- The addition of MTBE or ethanol appears to have only a small effect on the exhaust emissions of RFGs. The most substantial of these appears to be related to the emissions of CO and air toxics. Data from AQIRP suggest that ethanol-containing fuels lead to greater exhaust emissions of acetaldehyde than do fuels with MTBE, but less formaldehyde. Data from the California Ethanol Testing Program indicate that the exhaust emissions from vehicles using ethanol-containing fuels are about 10% lower than those arising from vehicles using fuels with MTBE. There is also some indication that oxygenates in fuels lead to somewhat higher emissions of NO_x—an effect that could have undesired impacts on air quality in rural areas and on regional scales.
- Ethanol-containing fuels tended to have significantly higher evaporative emissions (on both a total-mass basis and a reactivity-weighted basis) than MTBE-containing fuels. This is likely due, at least in part, to the fact that ethanol fuels tend to have an approximate 1-psi-higher RVP than the equivalent MTBE fuel. Moreover, the increase in the evaporative emissions from the ethanol-containing fuels was significantly larger than the slight benefit obtained from the lowering of the CO exhaust emissions using the ethanol-containing fuel.

- Based on the findings presented above, the committee concludes that the use of commonly available oxygenates in RFG has little impact on improving ozone air quality. Also, use of an ethanol-containing RFG with a 1-psi-higher RVP is likely to produce a negative air-quality impact. This conclusion is consistent with CARB's evaluation in 1998 that led to its decision to not allow a 1-psi waiver for ethanol-containing fuels (CARB 1998).

Two important caveats should be noted. The first relates to the fact that the analysis presented here is based solely on data gathered from well-maintained vehicles with properly working catalytic converters. As noted in Chapters 4 and 6, there is substantial evidence to suggest that high-emitting motor vehicles (perhaps because of malfunctioning catalytic converters or faulty evaporative controls) can contribute disproportionately to the VOC and CO emissions arising from a fleet of LDVs, and the response of high-emitting vehicles to ethanol-blended and MTBE-blended RFG has yet to be fully characterized. For example, one might speculate that oxygen in the fuel would provide a greater emissions benefit for high emitters with faulty catalytic converters than for ordinary vehicles. Because ethanol fuels often contain more oxygen than the equivalent MTBE fuel, this might tend to offset the disadvantages of ethanol-containing fuels implied in the committee's analysis. However, the few data on this subject that are currently available are inconclusive (e.g., see Knepper et al 1993; Mayotte et al. 1994). Moreover, for high-emitting vehicles with faulty evaporative controls, the use of ethanol-blended RFG with a higher RVP would most likely lead to elevated evaporative emissions. For these reasons, the committee recommends that the effect of RFG on emissions from high-emitting vehicles be studied in greater detail.

The other caveat relates to the overall effect on ozone pollution that might arise from the emissions differences projected here for MTBE-containing and ethanol-containing RFG blends. Recall from the committee's earlier analyses that the overall effect of RFGs might be an approximate 20% reduction in the reactivity of LDV emissions and a few parts-per-billion reduction in peak ozone concentrations. After combining exhaust and evaporative emissions, the use of ethanol, as opposed to MTBE, as an oxygenate would lead to a decrease in the effectiveness of RFGs but not a total cancellation. The net effect on ozone concentrations would be extremely small and almost certainly not discernable from the ambient ozone concentration data.

References

Anderson, L.G., P. Wolfe, R.A. Barrell, and J.A. Lanning. 1994. The effects of oxygenated fuels on the atmospheric concentrations of carbon monoxide and aldehydes in Colorado. Pp. 75-103 in Alternative Fuels and the Environment. F.S. Sterrett, ed. Boca Raton, FL.: Lewis Publishers.

Andersson-Skold, Y.P. Grennfelt, and K. Pleijel. 1992. Photochemical ozone creation potential: a study of different concepts. J. Air Waste Manage. Assoc. 42:1152-1158.

Andrews, T. 1874. Ozon: V. Ueber das ozon. Ann. Physik. Chem. 152:311-331.

AQIRP (Auto/Oil Air Quality Improvement Research Program). 1990. Initial Mass Exhaust Emission Results from Reformulated Gasolines. Technical Bulletin No. 1. Coordinating Research Council, Atlanta, GA.

AQIRP (Auto/Oil Air Quality Improvement Research Program). 1992. Effects of Fuel Sulfur on Mass Exhaust Emissions, Air Toxics, and Reactivity. Technical Bulletin No. 8. Coordinating Research Council, Atlanta, GA.

AQIRP (Auto/Oil Air Quality Improvement Research Program). 1993a. Reactivity Estimates for Reformulated Gasolines and Methanol/Gasoline Mixtures. Technical Bulletin No. 12, pp. 6-12a. Coordinating Research Council, Atlanta, GA.

AQIRP (Auto/Oil Air Quality Improvement Research Program). 1993b. Reactivity Estimates for Reformulated Gasolines and Methanol/Gasoline Blends in Prototype Flexible/Variable Fuel Vehicles. Technical Bulletin No. 7, Coordinating Research Council, Atlanta, GA.

AQIRP (Auto/Oil Air Quality Improvement Research Program). 1995. Gasoline Reformulation and Vehicle Technology Effects on Exhaust Emissions. Technical Bulletin No. 17. Coordinating Research Council, Atlanta, GA.

AQIRP (Auto/Oil Air Quality Improvement Research Program). 1997a. Predicted Effects of Reformulated Gasoline T50, T90, Sulfur and Oxygen

Content on Air Quality in Years 2000 and 2010. Technical Bulletin No. 21. Coordinating Research Council, Atlanta, GA.

AQIRP (Auto/Oil Air Quality Improvement Research Program). 1997b. Program Final Report. Auto/Oil Air Quality Improvement Research Program, Detroit, MI.

Ashbaugh, L.L., D.R. Lawson, G.A. Bishop, P.L. Guenther, D.H. Stedman, R.D. Stephens, P.J. Groblicki, J.S. Parikh, B.J. Johnson, and S.C. Huang. 1992. On-road remote sensing of carbon monoxide and hydrocarbon emissions during several vehicle operating conditions. Presented at Air & Waste Management Association/Environmental Particulate Source Controls, Phoenix, AZ.

Atkinson, R. 1994. Gas-phase tropospheric chemistry of organic compounds. J. Phys. Chem. Ref. Data Monog. 2:1-216.

Atkinson, R. 1997. Gas-phase tropospheric chemistry of volatile organic compounds: 1. Alkanes and alkenes. J. Phys. Chem. Ref. Data 26:215-290.

Atkinson, R. In press. Atmospheric oxidation. In Handbook of Property Estimation Methods for Chemicals: Environmental and Health Sciences. D. Mackay and R.S. Boethling, eds. Boca Raton, FL: CRC Press.

Beaton, S.P., G.A. Bishop, Y. Zhang, L.L. Ashbaugh, D.R. Lawson, and D.H. Stedman. 1995. On-road vehicle emissions: regulations, costs, benefits. Science 268(5213):991-993.

Bergin, M.S., A.G. Russell, and J.B. Milford. 1995. Quantification of individual VOC reactivity using a chemically detailed, three-dimensional photochemical model. Environ. Sci. Technol. 29(12):3029-3037.

Bergin, M.S., A.G. Russell, Y.J. Yang, J.B. Milford, F. Kirchner, and W.R. Stockwell. 1996. Effects of Uncertainty in SAPRC90 Rate Constants and Product Yields on Reactivity Adjustment Factors for Alternatively Fueled Vehicle Emissions. Report to the National Renewable Energy Laboratory, Golden, CO, 80401. NREL/TP-425-7636. July.

Bergin, M.S., A.G. Russell, W.P.L. Carter, B.E. Croes, and J.H. Seinfeld. 1998a. Ozone control and VOC reactivity. Pp. 3355-3383 in Encyclopedia of Environmental Analysis and Remediation. R.A. Meyers, ed. New York: John Wiley & Sons.

Bergin, M.S., A.G. Russell, and J.B. Milford. 1998b. Effects of chemical mechanism uncertainties on the reactivity quantification of volatile organic compounds using a three-dimensional air quality model. Environ. Sci. Technol. 32(5):694-703.

Bishop, G.A., and D.H. Stedman. 1989. Oxygenated Fuels, a Remote Sensing Evaluation (of CO Emission by Vehicles). SAE Technical Paper Series No. 891116. Paper for Government/Industry Meeting and Exposition May 1989, Washington, DC.

Bishop, G.A., and D.H. Stedman. 1990. On-road carbon monoxide emission measurement comparisons for the 1988-1989 Colorado (USA) oxy-fuels program. Environ. Sci. Technol. 24(6):843-847.

Bishop, G.A., and D.H. Stedman. 1995. Automobile emissions—control. Pp. 359-369 in Encyclopedia of Energy Technology and the Environment, Volume 1. A. Bisio and S. Boots, eds. New York: John Wiley & Sons.

Bishop, G.A., J.R. Starkey, A. Ihlenfeldt, W.J. Williams, and D.H. Stedman. 1989. IR long-path photometry: a remote sensing tool for automobile emissions. Anal. Chem. 61(10):671A-677A.

Bowman, F.M., and J.H. Seinfeld. 1994. Ozone productivity of atmospheric organics. J. Geophys. Res. 99(3):5309-5324.

Bowman, F.M., and J.H. Seinfeld. 1995. Atmospheric chemistry of alternate fuels and reformulated gasoline components. Prog. Energy Comb. Sci. 21(5):387-417.

Brooks, D.J., R.J. Peltier, S.L. Baldus, R.M. Reuter, W.J. Bandy III, and T.L. Sprik. 1995. Real-World Hot Soak Evaporative Emissions—A Pilot Study. SAE Paper 951007. Pp. 1-20 in Auto/Oil Air Quality Improvement Research Program, Vol. III. SP-1117. Society of Automotive Engineers, Warrendale, PA.

Burns, V.R., J.D. Benson, A.M. Hochhauser, W.J. Koehl, W.M. Dreucher, and R.M. Reuter. 1992. Description of the Auto/Oil Air Quality Improvement Research Program. SAE technical paper no. 912320. Pp. 1-28 in Auto/Oil Air Quality Improvement Research Program, SAE SP-920. Society of Automotive Engineers, Warrendale, PA.

Burns, V.R., A.M. Hochhauser, R.M. Reuter, L.A. Rapp, J.C. Knepper, B. Rippon, W.J. Koehl, W.R. Leppard, J.A. Rutherford, J.D. Benson, and L.J. Painter. 1995. Gasoline Reformulation and Vehicle Technology Effects on Emissions—Auto/Oil Air Quality Improvement Research Program. Pp. 143-165 in Auto/Oil Air Quality Improvement Research Program, Volume III. SAE SP-1117. Society of Automotive Engineers, Warrendale, PA.

Butler, J.W., C.A. Gierczak, G. Jesion, D.H. Stedman, and J.M. Lesko. 1994. On-road NOx Emissions Intercomparison of On-board Measurements and Remote Sensing. Final Report. CRC Report No. VE-11-6. Coordinating Research Council, Atlanta, GA.

Cal EPA (California Environmental Protection Agency). 1996. Comparison of Federal and California Reformulated Gasoline. Formerly California RFG Fact Sheet 3. Air Resources Board, Sacramento, CA. February.

Calvert, J.G., J.B. Heywood, R.F. Sawyer, and J.H. Seinfeld. 1993. Achieving acceptable air quality: some reflections on controlling vehicle emissions. Science 261:37-45.

Canale, R.P., S.R. Winegarden, C.R. Carlson, and D.L. Miles. 1978. General Motors Phase II Catalyst System. SAE Transactions 87:843-852.

CARB (California Air Resources Board). 1990. Proposed Regulations for Low-Emission Vehicles and Clean Fuels—Staff Report and Technical Support Document, Sacramento, CA., August 13. See also Appendix VIII of California Exhaust Emission Standards and Test Procedures for 1988 and Subsequent Model Passenger Cars, Light Duty Trucks and Medium Duty

Vehicles, as last amended September 22, 1993. Incorporated by reference in Section 1960.1 (k) of Title 13, California Code of Regulations.

CARB (California Air Resources Board). 1991. Proposed Reactivity Adjustment Factors for Transitional Low-Emissions Vehicles—Staff Report and Technical Support Document. Sacramento, CA. September 27.

CARB (California Air Resources Board). 1998. Proposed Determination Pursuant to Health and Safety Code Section 43830(g) of the Ozone Forming Potential of Elevated RVP Gasoline Containing 10 Percent Ethanol. California Environmental Protection Agency, California Air Resources Board, Sacramento, CA. November.

Cardelino, C.A., and W.L. Chameides. 1995. An observation-based model for analyzing ozone precursor relationships in the urban atmosphere. J. Air Waste Manage. Assoc. 45:161-180.

Carter, W.P.L. 1990. A detailed mechanism for the gas-phase atmospheric reactions of organic compounds. Atmos. Environ. Part A Gen Top. 24(3): 481-518.

Carter, W.P.L. 1993. Development and Application of an Up-to-Date Photochemical Mechanism for Airshed Modeling and Reactivity Assessment, Draft final report for California Air Resources Board Contract No. A934-094 Research Division, Sacramento, CA. April 26.

Carter, W.P.L. 1994. Development of ozone reactivity scales for Volatile Organic Compounds. J. Air Waste Manage. Assoc. 44:881-899.

Carter, W.P.L. 1995. Computer modeling environmental chamber measurements of maximum incremental reactivities of volatile organic compounds. Atmos. Environ. 29(18):2513-2517.

Carter, W.P.L. 1997. Summary of status of VOC reactivity estimates as of 12/10/97. [Online]. Available: http://www.cert.ucr.edu/~carter/rcttab.htm

Carter, W.P.L., and R. Atkinson. 1987. An experimental study of incremental hydrocarbon reactivity. Environ. Sci. Technol. 21(7):670-679.

Carter, W.P.L., and R. Atkinson. 1989. Computer modeling study of incremental hydrocarbon reactivity. Environ. Sci. Technol. 23(7):864-880.

Carter, W.P.L., and F.W. Lurmann. 1991. Evaluation of a detailed gas-phase atmospheric reaction mechanism using environmental chamber data. Atmos. Environ. Part A Gen. Top. 25(12):2771-2806.

Carter, W.P.L., J.A. Pierce, I.L. Malkina, D. Luo, and W.D. Long. 1993. Environmental Chamber Studies of Maximum Incremental Reactivities of Volatile Organic Compounds. Report to Coordinating Research Council, Project No. ME-9, California Air Resources Board Contract No. A032-0692; South Coast Air Quality Management District Contract No. C91323, U.S. Environmental Protection Agency Cooperative Agreement No. CR-814396-01-0, University Corporation for Atmospheric Research Contract No. 59166, and Dow Corning Corporation. April 1.

Carter, W.P.L., D. Luo, I.L. Malkina, and D. Fitz. 1995a. The University of California, Riverside Environmental Chamber Data Base for Evaluating

Oxidant Mechanism. Indoor Chamber Experiments through 1993. Report submitted to the U. S. Environmental Protection Agency, EPA/AREAL, Research Triangle Park, NC. March 20.

Carter, W.P.L., J.A. Pierce, D. Luo, and I.L. Malkina. 1995b. Environmental chamber study of maximum incremental reactivities of Volatile Organic Compounds. Atmos. Environ. 29(18):2499-2511.

Carter, W.P.L., D. Luo, I.L. Malkina, and J.A. Pierce. 1995c. Environmental Chamber Studies of Atmospheric Reactivities of Volatile Organic Compounds. Effects of Varying ROG Surrogate and NOx. Final report to Coordinating Research Council, Project ME-9; California Air Resources Board, Contract A032-0692; and South Coast Air Quality Management District, Contract C91323. March 24.

Chameides, W.L., F. Fehsenfeld, M.O. Rodgers, C. Cardelino, J. Martinez, D. Parrish, W. Lonneman, D.R. Lawson, R.A. Rasmussen, P. Zimmerman, J. Greenburg, P. Middleton, and T. Wang. 1992. Ozone precursor relationships in the ambient atmosphere. J. Geophys. Res. 97:6037.

Chameides, W.L., R.D. Saylor, and E.B. Cowling. 1997. Ozone pollution in the rural United States and the new NAAQS. Science 276:916.

Chang, T.Y., S.J. Rudy, G. Kuntasal, and R.A. Gorse, Jr. 1989. Impact of methanol vehicles on ozone air quality. Atmos. Environ. 23(8):1629-1644.

Cleary, K. 1998. EXCEL spreadsheet furnished by the California Air Resources Board, Sacramento, CA. May.

Cook, R., P. Enns, and M.S. Sklar. 1997. Regression Analysis of Ambient CO Data from Oxyfuel and Nonoxyfuel Areas. Paper No. 97-RP139.02. Air and Waste Management Association 90th Annual Meeting & Exhibition, June 8-13, 1997, Toronto, Canada.

Cox, W.M., and S.-H. Chu. 1993. Meteorologically adjusted ozone trends in urban areas: a probabilistic approach. Atmos. Environ. Pt. B, Urban Atmos. 27(4):425-434.

Darlington, T.L., P.E. Korsog, and R.S. Strassburger. 1992. Real World Vehicle and Engine Operations: Results of the MVMA/AIAM Instrumented Vehicle Pilot Study. Paper No. 92-1075 presented at the 85th Annual Meeting and Exhibition of the Air & Waste Management Association, June 21-26, 1992, Kansas City, MO.

Darnall, K.R., A.C. Lloyd, A.M. Winer, and J.N. Pitts, Jr. 1976. Reactivity scale for atmospheric hydrocarbons based on reaction with hydroxyl radical. Environ. Sci. Technol. 10(7):692-696.

Davis, S.C. 1997. Transportation Energy Data Book: Edition 17. ORNL-6919. Center for Transportation Analysis, Oak Ridge National Laboratory, prepared for U.S. Department of Energy, Washington, DC.

Davis, S.C. 1998. Transportation Energy Data Book: Edition 18. ORNL-6941. Center for Transportation Analysis, Oak Ridge National Laboratory, prepared for U.S. Department of Energy, Washington, DC.

Dennis, R.L., D.W. Byun, J.H. Novak, K.J. Galluppi, C.J. Coats, and M.A. Vouk.

1996. The next generation of integrated air quality modeling: EPA's models-3. Atmos. Environ. 30(12):1925-1938.

Derwent, R.G., and A.M. Hov. 1979. Computer Modeling Studies of Photochemical Air Pollution Formation in North West Europe. AERE R 9434, Harwell Laboratory, Oxfordshire, England.

Derwent, R.G., and O. Hov. 1988. Application of sensitivity and uncertainty analysis techniques to a photochemical ozone model. J. Geophys. Res. 93(5):5185-5199.

Derwent, R.G., and M.E. Jenkin. 1991. Hydrocarbons and the long range transport of ozone and PAN across Europe. Atmos. Environ. Part A Gen. Top. 25(8):1661-1673.

Derwent, R.G., M.E. Jenkin, and S.M. Saunders. 1996. Photochemical ozone creation potentials for a large number of reactive hydrocarbons under European conditions. Atmos. Environ. 30:181-199.

Derwent, R.G., M.E. Jenkin, S.M. Saunders, and M.J. Pilling. 1998. Photochemical ozone creation potentials for organic compounds in Northwest Europe calculated with a master chemical mechanism. Atmos. Environ. 32:2429-2441.

Dimitriades, B. 1996. Scientific basis for the VOC reactivity issues raised by Section 193(3) of the Clean Air Amendments of 1990. J. Air Waste Manage. Assoc. 46(10):963-970.

Dodge, M.C. 1977. Combined use of modeling techniques and smog chamber data to derive ozone–precursor relationships. Pp. 881-889 in International Conference on Photochemical Oxidant Pollution and its Control: Proceedings, Vol. II, B. Dimitriades, ed. EPA/600/3/77-001b. U.S. Environmental Protection Agency, Environmental Sciences Research Laboratory, Research Triangle Park, NC.

Dodge, M.C. 1984. Combined effects of organic reactivity and NMHC/NOx ratio on photochemical oxidant formatiom—A modeling study. Atmos. Environ. 18:1657.

Dodge, M.C. In press. Chemical oxidant mechanisms for air quality modeling. Atmos. Environ.

Dolislager, L.J. 1997. The effect of California's wintertime oxygenated fuels program on ambient carbon monoxide concentrations. J. Air Waste Manage. Assoc. 47(7):775.

Dunker, A.M., R.E. Morris, A.K. Pollack, C.H. Schleyer, and G. Yarwood. 1996. Photochemical modeling of the impact of fuels and vehicles on urban ozone using auto/oil program data. Environ. Sci. Technol. 30(3):787-801.

EPA (U.S. Environmental Protection Agency). 1973. Regulation of Fuels and Fuel Additives. Fed. Regist. 38(6):1253-1261.

EPA (U.S. Environmental Protection Agency). 1980. Passenger Car Fuel Economy: EPA and Road. EPA 460/3-80-010. Office of Mobile Sources Air Pollution Control, Emission Control Technology Division, Ann Arbor, MI.

EPA (U.S. Environmental Protection Agency). 1989. Volatility Regulations for

Gasoline and Alcohol Blends Sold in Calendar Years 1989 and Beyond. Fed. Regist. 54(54):11868-11911.

EPA (U.S. Environmental Protection Agency). 1990. Volatility Regulations for Gasoline and Alcohol Blends Sold in Calendar Years 1992 and Beyond. Fed. Regist. 55(112):23658-23667.

EPA (U.S. Environmental Protection Agency). 1994. Regulation of Fuels and Fuel Additives: Standards for Reformulated and Conventional Gasoline, Final Rule. Fed. Regist. 59(32):7716-7878.

EPA (U.S. Environmental Protection Agency). 1996a. National Air Quality and Emissions Trends Report, 1995. EPA 454/R-97-013. U.S. Environmental Protection Agency, Washington, DC.

EPA (U.S. Environmental Protection Agency). 1996b. Final Regulations for Revisions to the Federal Test Procedure for Emissions from Motor Vehicles. 40 CFR Part 86, Fed. Regist. 61(205):54851-55906.

EPA (U.S. Environmental Protection Agency). 1997a. National Ambient Air Quality Standards for Ozone; Final Rule. Fed. Regist. 62(138):1-37.

EPA (U.S. Environmental Protection Agency). 1997b. Fuels and Fuel Additives; Elimination of Oxygenated Fuels Program Reformulated Gasoline (OPRG) Category From the Reformulated Gasoline Regulations. Final Rule. Fed. Regist. 62(Nov. 6):60131-60136.

EPA (U.S. Environmental Protection Agency). 1998. National Air Quality and Emissions Trends Report, 1997. EPA 454/R-98-016. U.S. Environmental Protections Agency, Office of Air Quality Planning and Standards, Research Triangle Park, NC.

FHWA (Federal Highway Administration). 1997. 1995 Nationwide Personal Transportation Survey: User's Guide for the Public Use Data Files. FHWA-PL-98-002. U.S. Department of Transportation, Washington, DC. October.

GAO (General Accounting Office). 1997. Air Pollution: Limitations of EPA's Motor Vehicle Emissions Model and Plans to Address Them. GAO/RCED-97-210. General Accounting Office, Washington, DC. September.

Gery, M.W., G.Z. Whitten, and J.P. Killus. 1988. Development and Testing of the CBM-IV for Urban and Regional Modeling. EPA-600/3-88-012. U.S. Environmental Protection Agency, Washington, DC. January.

Gery, M.W., G.Z. Whitten, J.P. Killus, and M.C. Dodge. 1989. A photochemical kinetics mechanism for urban and regional scale computer modeling. J. Geophys. Res. 94:12925-12956.

Gipson, G.L. 1984. User's Manual for OZIPM-2: Ozone Isopleth Plotting with Optional Mechanisms/Version 2. EPA/450/4-84-024. U.S. Environmental Protection Agency, Washington, DC. August.

Guenther, P.L., D.H. Stedman, G.A. Bishop, J. Hannigan, J. Bean, and R. Quine. 1991. Remote sensing of automobile exhaust. API Publication No. 4538. Washington, D.C.: American Petroleum Institute.

Guthrie, P.D., M.P. Ligocki, R.E. Looker, and J.P. Cohen. 1996. Air Quality Effects of Alternative Fuels. Phase 1. Draft Final Report for Subcontract

14072-02 to the National Renewable Energy Laboratory, Golden, CO. December.

Haagen-Smit, A.J. 1952. Chemistry and physiology of Los Angeles smog. Ind. Eng. Chem. 44:1342-1346.

Haagen-Smit, A.J., and M.M. Fox. 1954. Photochemical ozone formation with hydrocarbons and automobile exhaust. J. Air Pollut. Control Assoc. 4:105-109.

Haagen-Smit, A.J., and M.M. Fox. 1955. Automobile exhaust and ozone formation. SAE Technical Series No. 550277. Society of Automotive Engineers, Warrendale, PA.

Haagen-Smit, A.J., and M.M. Fox. 1956. Ozone formation in photochemical oxidation of organic substances. Ind. Eng. Chem. 48:1484-.

Haagen-Smit, A.J., E.F. Darley, M. Zaitlin, H.Hull, and W. Noble. 1951. Investigation on injury to plants from air pollution in the Los Angeles area. Plant Physiol. 27:18-.

Haagen-Smit, A.J., C.E. Bradley, and M.M. Fox. 1953. Ozone formation in photochemical oxidation of organic substances. Ind. Eng. Chem. 45:2086.

Harley, R.A., A.G. Russell, G.J. McRae, L.A. McNair, and D.A. Winner. 1992. Continued Development of a Photochemical Model and Application to the Southern California Air Quality Study (SCAQS) Intensive Monitoring Periods: Phase 1. GRA&I Issue 19.

Harley, R.A., A.G. Russell, G.J. McRae, G.R. Cass, and J.H. Seinfeld. 1993. Photochemical modeling of the Southern California Air Quality Study. Environ. Sci. Technol. 27(2):378-388.

Harley, R.A., R.F. Sawyer, and J.B. Milford. 1997. Updated photochemical modeling for California's South Coast Air Basin: Comparison of chemical mechanisms and motor vehicle emission inventories. Environ. Sci. Technol. 31(10):2829-2839.

Heywood, J.B. 1988. Internal Combustion Engine Fundamentals. New York: McGraw-Hill.

Hochhauser, A.M., J.D. Benson, V.R. Burns, R.A. Gorse, Jr., W.J. Koehl, L.J. Painter, R.M. Reuter, and J.A. Rutherford. 1992. Speciation and Calculated Reactivity of Automotive Exhaust Emissions and Their Relation to Fuel Properties—Auto/Oil Air Quality Improvement Research Program. Technical paper 920325. Pp. 359-390 in Auto/Oil Air Quality Improvement Research Program, SP-920. Society of Automotive Engineers, Warrendale, PA.

Hogrefe, C., S.T. Rao, and I.G. Zurbenko. 1998. Detecting trends and biases in time series of ozonesonde data. Atmos. Environ. 32(14/15):2569-2586.

Jeffries, H.E., and K.G. Sexton. 1995. The Relative Ozone Forming Potential of Methanol-Fueled Vehicle Emissions and Gasoline-Fueled Vehicle Emissions in Outdoor Smog Chambers. Final Report to the Coordinating Research Council, Atlanta, GA, for Project No. ME-1. January 1995.

Jeffries, H., K. Sexton, C. Rigsbee, and W. Li. 1997. Experimental and Theoretical Studies of Reactivity as a Conserved Property Using an Airtrak. GRA&I Issue 24.

Jeffries, H., K. Sexton, and J. Yu. 1998. Atmospheric Photochemical Studies of Pollutant Emissions from Transportation Vehicles Operating on Alternative Fuels. National Renewable Energy Laboratory, Golden, CO. June.

Jenkin, M.E., S.M. Saunders, and M.J. Pilling. 1997. The tropospheric degradation of volatile organic compounds—A protocol for mechanism development. Atmos. Environ. 31:81-104.

Jiang, W., D.L. Singleton, M. Hedley, and R. McLaren. 1996. Sensitivity of ozone concentrations to VOC and NOx emissions in the Canadian Lower Fraser Valley. Atmos. Environ. 31(4):627-638.

Kelly, N.A., and P.J. Groblicki. 1992. Real-World Emissions from a Modern Production Vehicle Driven in Los Angeles. General Motors Research Publication No. GMR-7858 EV-403. General Motors, Warren, MI. December 9.

Kelly, N.A., P. Wang, S.M. Japar, M.D. Hurley, and T.J. Wallington. 1994. Measurement of the Atmospheric Reactivity of Emissions from Gasoline and Alternative-Fueled Vehicles—Assessment of Available Methodologies. Part 1. Indoor Smog Chamber Study of Reactivity. Final Report for the First Year, January 1-December 31, 1993. Environmental Research Consortium, Warren, MI; General Motors Research and Development Center, Warren, MI; Environmental Research Dept., Ford Motor Co., Dearborn, MI; Scientific and Research Lab., Coordinating Research Council, Atlanta, GA; National Renewable Energy Lab., Golden, CO.

Kelly, N.A., P. Wang, M.D. Hurley, S.M. Japar, T. Chang, and T.J. Wallington. 1996. Measurement of the Atmospheric Reactivity of Emissions from Gasoline and Alternative-Fueled Vehicles—Assessment of Available Methodologies. Part 1. Indoor Smog Chamber Study of Reactivity. Part 2. Assessment of AirTrak as a Reactivity Analyzer. Final Report for the Second Year, March 1, 1994-February 28, 1995. Environmental Research Consortium, Warren, MI; Designers Diversified Services, Beverly Hills, MI; Coordinating Research Council, Atlanta, GA; National Renewable Energy Lab., Golden, CO.

Khan, M., Y.J. Yang, and A.G. Russell. 1999. Photochemical reactivities of common solvents: Comparison between urban and regional domains. Atmos. Environ. 33(7):1085-1092.

Kirchstetter, T.W., B.C. Singer, R.A. Harley, G.R. Kendall, and W. Chan. 1996. Impact of oxygenated gasoline use on California light-duty vehicle emissions. Environ. Sci. Technol. 30(2):661-670.

Kirchstetter, T.W., R.A. Harley, G.R. Kendall, and J. Hesson. 1997. Impact of California Phase 2 Reformulated Gasoline on Atmospheric Reactivity of Exhaust and Evaporative Emissions. Air and Waste Management Association 90th Annual Meeting and Exhibition, June 8-13, 1997. Toronto, Ontario, Canada.

Kirchstetter, T.W.. B.C. Singer, R.A. Harley, G.R. Kendall, and M. Traverse.

1999a. Impact of California reformulated gasoline on motor vehicle emissions: 1. Mass emission rates. Environ. Sci. Technol. 33(2):318-328.

Kirchstetter, T.W., B.C. Singer, R.A. Harley, G.R. Kendall, and J.M. Hesson. 1999b. Impact of California reformulated gasoline on motor vehicle emissions: 2. Volatile organic compound speciation and reactivity. Environ. Sci. Technol. 33(2):329-336.

Kishan, S., T.H. DeFries, and C.G. Weyn. 1993. A study of light-duty vehicle driving behavior: application to real-world emission inventories. SAE Technical Paper 932659. Pp. 1-15 in Advanced Emission Control Technologies, SP-997. Society of Automotive Engineers, Warrendale, PA.

Kleindienst, T.E., J.A. Stikeleather, C.O. Davis, D. Smith, and F. Black. 1996. Photochemical Reactivity Studies of Emissions from Alternative Fuels. Report to the National Renewable Energy Laboratory for Contract TS-2-1116-1 and to the Atmospheric Research and Exposure Assessment Laboratory of the U.S. Environmental Protection Agency, National Renewable Energy Laboratory, Golden, CO. March.

Kleinman, L.J. 1994. Low and high NO_x tropospheric photochemistry. J. Geophys. Res. 99:16831-16838.

Kleinman, L.J. In press. Observation based analysis for ozone production. Atmos. Environ.

Knepper, J.C., W.J. Koehl, J.D. Benson, V.R. Burns, R.A. Gorse, Jr., A.M. Hochhauser, W.R. Leppard, L.A. Rapp, and R.M. Reuter. 1993. Fuel effects in auto/oil high emitting vehicles. Pp. 1-17 in Auto/Oil Air Quality Improvement Research Program, Volume II. SP-1000. Society of Automotive Engineers, Warrendale, PA.

Kwok, E.S.C., and R. Atkinson. 1995. Estimation of hydroxyl radical reaction rate constants for gas-phase organic compounds using a structure-reactivity relationship: an update. Atmos. Environ. 29(14):1685-1695.

Lamb, R.G. 1983. Regional Scale (1000 km) Model of Photochemical Air Pollution, Part 1. Theoretical Formulation. EPA/600/3-83-035. U.S. Environmental Protection Agency, Environmental Sciences Research Laboratories, Research Triangle Park, NC.

Larsen, L.C., and S.J. Brisby. 1998. Cleaner-Burning Gasoline: An Assessment of Its Impact on Ozone Air Quality in California. California Environmental Protection Agency, California Air Resources Board, Sacramento, CA. January.

Lawson, D.R., P.J. Groblicki, D.H. Stedman, G.A. Bishop, and P.L. Guenther. 1990. Emissions from in-use motor vehicles in Los Angeles: A pilot study of remote sensing and the inspection and maintenance program. J. Air Waste Manage. Assoc. 40(8):1096-1105.

Le Calve, S., G. Le Bras, and A. Mellouki. 1997. Temperature dependence for the rate coefficients of the reactions of the OH radical with a series of formates. J. Phys. Chem. A 101(30):5489-5493.

Leighton, P.A. 1961. Photochemistry of Air Pollution. New York: Academic Press.

Lurmann, F.W., W.P.L. Carter, and R.A. Coyner. 1987. A Surrogate Species Chemical Reaction Mechanism for Urban-Scale Air Quality Simulation Models. Volume I. Adaptation of the Mechanism. EPA-600/3-87-014a. U.S. Environmental Protection Agency, Research Triangle Park, NC. June.

Mannino, D.M., and R.A. Etzel. 1996. Are oxygenated fuels effective? An evaluation of ambient carbon monoxide concentrations in 11 western states, 1986-1992. J. Air Waste Manage. Assoc. 46:20-24.

Mayotte, S.C., C.E. Lindhjem, V. Rao, and M.S. Sklar. 1994. Reformulated Gasoline Effects on Exhaust Emissions: Phase I: Initial Investigation of Oxygenate, Volatility, Distillation and Sulfur Effects. SAE Technical Paper Series 941973. Society of Automotive Engineers, Warrendale, PA.

McBride, S.J., M.A. Matthew, and A.G. Russell. 1997. Cost–benefit and uncertainty issues in using organic reactivity to regulate urban ozone. Environ. Sci. Technol. 31(5):238A-244A.

McNair, L., A. Russell, and M.T. Odman. 1992. Airshed calculation of the sensitivity of pollutant formation to organic compound classes and oxygenates associated with alternative fuels. J. Air Waste Manage. Assoc. 42: 174-178.

McNair, L.A., A.G. Russell, M.T. Odman, B.E. Croes, and L. Kao. 1994. Airshed model evaluation of reactivity adjustment factors calculated with the maximum incremental reactivity scale for transitional low emission vehicles. J. Air Waste Manage. Assoc. 44(7):900-907.

McRae, G.J., W.R. Goodin, and J.H. Seinfeld. 1982. Mathematical Modeling of Photochemical Air Pollution. EQL Report No. 18. Final Report to the State of California Air Resources Board, Sacramento, CA.

Middleton, J.T., J.B. Kendrick, and H.W. Schwalm. 1950. Injury to herbaceous plants by smog or air pollution. Plant Disease Reptr. 34:245-252.

Milanchus, M.L., S.T. Rao, and I.G. Zurbenko. 1998. Evaluating the effectiveness of ozone management efforts in the presence of meteorological variability. J. Air Waste Manage. Assoc. 48:201-215.

NAPAP (U.S. National Acid Precipitation Assessment Program). 1991. Acidic Deposition: State of Science and Technology, Vol. IV, pp. 25-390. Control Technologies, Future Emissions, and Effects Valuation. Washington, DC: U.S. Government Printing Office.

NRC (National Research Council). 1991. Pp. 31-39 and 413-424 in Rethinking the Ozone Problem in Urban and Regional Air Pollution. Washington, D.C.: National Academy Press.

NRC (National Research Council). 1996. Toxicological and Performance Aspects of Oxygenated Motor Vehicle Fuels. Washington, D.C.: National Academy Press.

NSTC (National Science and Technology Council). 1997. Interagency Assessment of Oxygenated Fuels. National Science and Technology Council, Committee on Environment and Natural Resources. Office of Science and Technology Policy, Executive Office of the President of the United States.

Odman, M.T., A. Xiu, and D.W. Byun. 1994. Evaluating advection schemes for use in the next generation of air quality modeling systems. GRA&I Issue 07.

OTA (Office of Technology Assessment). 1990. Replacing Gasoline: Alternative Fuels for Light-Duty Vehicles. Washington, D.C.: U.S. Government Printing Office.

OTAG (Ozone Transport Assessment Group). 1997. Telling the OTAG Story With Data: Final Report. Vol. 1: Executive Summary [Online]. Available: http://capita.wustl.edu/OTAG/Reports/AQAFinVol_I/HTML/v1_exsum7.html.

PRC (PRC Environmental Management, Inc.). 1992. Performance Audit of the Colorado Oxygenated Fuels Program. Colorado State Auditor, State of Colorado, Denver, CO.

Rao, S.T., G. Sistla, and R. Henry. 1992. Statistical analysis of trends in urban ozone air quality. J. Air Waste Manage. Assoc. 42(9):1204-1211.

Rao, S.T., I.G. Zurbenko, R. Neagu, P.S. Porter, J.Y. Ku, and R.F. Henry. 1997. Space and time scales in ambient ozone data. Bull. Am. Meteor. Soc. 70:2153-2166.

Rao, S.T., P.S. Porter, I.G. Zurbenko, A.M. Dunker, and G.T. Wolff. 1998. Ozone Air Quality Over North America: A Critical Review of Trend Detection Techniques and Assessments. NARSTO Synthesis Team Draft: 12/03/98 [Online]. Available: http://www.cgenv.com/Narsto/

Reid, R.C., J.M. Prausnitz, and B.E. Poling. 1987. The Properties of Gases and Liquids, 4th Ed. New York: McGraw-Hill.

Reuter, R.M., R.A. Gorse, Jr., L.J. Painter, J.D. Benson, A.M. Hochhauser, B.H. Rippon, V.R. Burns, W.J. Koehl, and J.A. Rutherford. 1992. Effects of oxygenated fuels and RVP on automotive emissions. Auto/Oil Air Quality Improvement Research Program, Technical Paper No. 920326. Pp. 391-412 in Auto/Oil Air Quality Improvement Research Program, SAE SP-920. Society of Automotive Engineers, Warrendale, PA. February.

Reynolds, S.D., P.M. Roth, and J.H. Seinfeld. 1973. Mathematical modeling of photochemical air pollution. I. Formulation of the model. Atmos. Environ. 7:1033-1061.

Reynolds, S.D., T.W. Tesche, and L.E. Reid. 1979. An Introduction to the SAI Airshed Model and its Usage. Report No. SAI-EF79-31. Systems Application Inc., San Rafael, CA.

Ross, M., R. Goodwin., R. Watkins, M.Q. Wang, and T. Wenzel. 1995. Real-World Emissions from Model Year 1993, 2000, and 2010 Passenger Cars. American Council for an Energy-Efficient Economy, Washington, DC.

Russell, A.G., and R.L. Dennis. 1998. NARSTO Critical Review of Photochemical Models and Modeling. NARSTO Synthesis Team Draft: 12/03/98. [Online]. Available: http://www.cgenv.com/Narsto/

Russell, A.G., D. St. Pierre, and J.B. Milford. 1990. Ozone control and methanol fuel use. Science 247(4939):201-205.

Russell, A., J. Milford, M.S. Bergin, S. McBride, L. McNair, Y. Yang, W.R. Stockwell, and B. Croes. 1995. Urban ozone control and atmospheric reactivity of organic gases. Science 269:491-495.

Sawyer, R.F., R.A. Harley, S.H. Cadle, H.A. Bravo, R. Slott, R.A. Gorse, and J.M. Norbeck. 1998. Mobile Sources Critical Review. NARSTO Synthesis Team Draft: 12/03/98 [Online]. Available: http://www.cgenv.com/Narsto

Schönbein, C.F. 1840. Beobachtungen über den bei der elektrolysation des wassers und dem Ausströmen der gewöhnliehen electricität aus spitzen sich entwikkelnden geruch. Ann. Phys. Chem. 50:616-635.

Schönbein, C.F. 1854. Über verschiedene zustände des sauerstoffs. Liebigs Ann. Chem. 89:257-300.

Seinfeld, J.H. 1986. Pp. 36-42 in Atmospheric Chemistry and Physics of Air Pollution. New York: John Wiley & Sons.

Stedman, D.H., G.A. Bishop, S.P. Beaton, J.E. Peterson, P.L. Guenther, I.F. McVey, and Y. Zhang. 1994. On-road Remote Sensing of CO and HC Emissions in California. Final Report to California Air Resources Board, Contract No. A032-093, Sacramento, CA.

Stedman, D.H., G.A. Bishop, P. Aldrete, and R.S. Slott. 1997. On-road evaluation of an automobile emission test program. Environ. Sci. Technol. 31:927-931.

Stephens, R.D., and S.H. Cadle. 1991. Remote sensing measurements of carbon monoxide emissions from on-road vehicles. J. Air Waste Manage. Assoc. 41:39-46.

Stephens, R.D., M. Giles, K. McAlinden, R.A. Gorse, Jr., D. Hoffman, and R. James. 1997. An analysis of Michigan and California CO remote sensing measurements. J. Air Waste Manage. Assoc. 47:601-607.

Stockwell, W.R., P. Middleton, J.S. Chang, and X. Tang. 1990. The second generation regional acid deposition model chemical mechanism for regional air quality modeling. J. Geophys. Res. 95(10):16343-16376.

Trijonis, J.C., and K.W. Arledge. 1976. Utility of Reactivity Criteria in Organic Emission Control Strategies: Application to the Los Angeles Atmosphere. EPA/600/3-78/019. TRW/Environmental Services, Redondo Beach, CA.

Weaver, C.S., and L.M. Chan. 1997. Comparison of Off-cycle and Cold-start Emissions from Dedicated NGVS and Gasoline Vehicles: Final Report. June 1995-August 1996. GRI-96/0217. Prepared by Engine, Fuel, and Emissions Engineering, Inc., Sacramento, CA. for Gas Research Institute, Natural Gas Vehicle Business Unit, Chicago, IL.

Whitten, G.Z., J.P. Cohen, and A.M. Kuklin. 1997. Regression Modeling of

Oxyfuel Effects on Ambient CO Concentrations: Final Report. SYSAPP-96/97. Prepared for Renewable Fuels Association and Oxygenated Fuels Association by System Applications International, Inc. San Rafael, CA. January.

Wolff, G.T. 1996. The scientific basis for a new ozone standard. Environ. Manager (September):27-32.

Yamartino, R.J., J.S. Scire, S.R. Hanna, G.R. Carmichael, and Y.S. Chang. 1989. CALGRID: A Mesoscale Photochemical Grid Model. Volume f 1: Model Formulation Document. Sigma Research Corporation, A049-1. Final report on Contract A6-215-74 for the California Air Resources Board, Sacramento, CA.

Yamartino, R.J., J.S. Scire, G.R. Carmichael, and Y.S. Chang. 1992. The CALGRID mesoscale photochemical grid model. 1. Model formulation. Atmos. Environ. Part A Gen. Top. 26(8):1493-1512.

Yang, Y.-J., and J.B. Milford. 1996. Quantification of uncertainty in reactivity adjustment factors from reformulated gasolines and methanol fuels. Environ. Sci. Tech. 30(1):196-203.

Yang, Y.-J., M. Das, J.B. Milford, M.S. Bergin, A.G. Russell, and W.R. Stockwell. 1994. Quantification of Organic Compound Reactivities and Effects of Uncertainties in Rate Parameters. An Integrated Approach Using Formal Sensitivity and Uncertainty Analysis and Three Dimensional Air Quality Modeling. Report prepared for the Auto/Oil Air Quality Improvement Research Program. Society of Automotive Engineers, Warrendale, PA. August.

Yang, Y.-J., W.R. Stockwell, and J.B. Milford. 1995. Uncertainties in incremental reactivities of volatile organic compounds. Environ. Sci. Technol. 29(5):1336-1345.

Yang, Y.-J., W.R. Stockwell, and J.B. Milford. 1996. Effect of chemical product yield uncertainties on reactivities of VOCs and emission from reformulated gasolines and methanol fuels. Environ. Sci. Technol. 30(4):1392-1397.

Zhang, J., and S.T. Rao. In press. On the role of vertical mixing in the temporal evolution of the ground-level ozone concentrations. J. Appl. Meteor.

Zhang, Y., D.H. Stedman, G.A. Bishop, P.L. Guenther, S.P. Beaton, and J.E. Peterson. 1993. On-road hydrocarbon remote sensing in the Denver area. Environ. Sci. Technol. 27(9):1885-1891.

Zhang, Y., G.A. Bishop, and D.H. Stedman. 1994. Automobile emissions are statistically gamma-distributed. Environ. Sci. Technol. 28(7):1370-1374.

Zhang, Y., D.H. Stedman, G.A. Bishop, S.P. Beaton, P.L. Guenther, and I.F. McVey. 1996a. Enhancement of remote sensing for mobile source nitric oxide. J. Air Waste Manage. Assoc. 46:25-29.

Zhang, Y., D.H. Stedman, G.A. Bishop, S.P. Beaton, and P.L. Guenther. 1996b.

On-road evaluation of inspection/maintenance effectiveness. Environ. Sci. Technol. 30:1445-1450.

Zurbenko, I.G., P.S. Porter, S.T. Rao, J.Y. Ku, R. Gui, and R.E. Eskridge. 1996. Detecting discontinuities in time series of upper air data: Development and demonstration of an adaptive filter technique. J. Climate 9(12): 3548-3560.

Appendix A

Biographical Information on the Committee on Ozone-Forming Potential of Reformulated Gasoline

William L. Chameides is Regents Professor of Earth and Atmospheric Sciences at the Georgia Institute of Technology. His research interests include atmospheric chemistry, tropospheric gas-phase and aqueous-phase chemistry; air pollution; global chemical cycles; biospheric-atmospheric interaction; and global and regional environmental change. His NRC service includes being the chair of the Committee on Atmospheric Chemistry and a member of the committee on Tropospheric Ozone Formation and Measurement. Dr. Chameides has a B.A. degree from SUNY Binghamton, and an M.Ph. and a Ph.D. in geology and geophysics from Yale University. He is a member of the National Academy of Sciences.

Charles A. Amann is head of KAB Engineering. Previously, he was a research fellow with General Motors Research Laboratories. Mr. Amann's research interests include fuels and combustion; internal combustion engines; and energy technologies. He received a B.S. and M.S. in mechanical engineering from the University of Minnesota. Mr. Amann is a member of the National Academy of Engineering and has served on many committees of the National Research Council.

Roger Atkinson is a research chemist at the University of California at

Riverside. His research areas include kinetics and mechanisms of the gas phase of the atmospherically important reactions of organic compounds. Dr. Atkinson received a B.A. in natural sciences and a Ph.D. in physical chemistry from Cambridge University. He currently serves as a member of the California Air Resources Board's Reactivity Scientific Advisory Committee. He has also served as a member of the NRC committee on Tropospheric Ozone Formation and Measurement.

Nancy J. Brown is a senior scientist at the Ernest Orlando Lawrence Berkeley National Laboratory. She is also head of the Environmental Research Program within the Energy and Environment Division, and leader of the Combustion Research Group. Dr. Brown's research areas include combustion chemistry, chemical dynamics, modeling combustion processes, and model sensitivity and uncertainty in atmospheric chemistry. She received a B.S. from Virginia Polytechnic Institute, and an M.S. and a Ph.D. in chemical physics from the University of Maryland.

Jack G. Calvert is a senior scientist at the National Center for Atmospheric Research. His research areas include atmospheric chemistry, photochemistry, reaction kinetics, and formation and decay mechanisms of reactive transients (molecules and free radicals). Dr. Calvert has served on many NRC committees and currently serves on the California Air Resources Board's Reactivity Scientific Advisory Committee. He received a B.S. and Ph.D. in chemistry from UCLA.

Fred C. Fehsenfeld is senior scientist and program leader for the National Oceanic and Atmospheric Administration (NOAA) Aeronomy Laboratory. His research areas include atmospheric measurement of ozone and precursors. Dr. Fehsenfeld received a B.A. from Rice University and a Ph.D. in physics from the University of Texas. He served as a member of the NRC Committee on Tropospheric Ozone Formation and Measurement and a member of the NRC Committee on Atmospheric Chemistry.

John P. Longwell is a professor emeritus in the Department of Chemical Engineering at the Massachusetts Institute of Technology. Previously, Dr. Longwell conducted research at Exxon Corporation on petroleum and petrochemicals and directed its Central Basic Research Laboratory. Dr. Longwell's relevant research interests are in fuels and petrochemicals, petroleum refining, environmental control, and fuel combustion. He is a member of the National Academy of Engineering and has served on many NRC committees. He received a B.S. from the University of California at Berkeley and a Sc.D. in chemical engineering at MIT.

Mario J. Molina is a professor of atmospheric chemistry in the Department of Earth, Atmospheric, and Planetary Sciences at Massachusetts Institute of Technology. His main research interests are in chemistry of the stratosphere. He was co-discoverer of the theory that chlorofluorocarbon gases deplete the ozone layer of the stratosphere. Dr. Molina is a Nobel Laureate and a member of the National Academy of Sciences. He received a B.S. from the University of National Autonoma De Mexico and a Ph.D. in physical chemistry from the University of California at Berkeley.

S. Trivikrama Rao is assistant commissioner for the Office of Science and Technology of the New York State Department of Environmental Conservation. He is also a research professor of atmospheric science in the Department of Earth and Atmospheric Sciences and professor of environmental statistics in the Department of Biometry and Statistics at the State University of New York (SUNY) at Albany. Dr. Rao's research areas include automobile pollution dispersion; atmospheric turbulence and air pollution meteorology; modeling of photochemical oxidants; and analysis and interpretation of environmental data. He received a B.Sc. from Andhra Loyola College in India, a M.Sc. (tech.) from Andhra University in India, and a Ph.D. in atmospheric science at SUNY Albany. He currently serves on the editorial board for the Atmospheric Environment Journal. Also, he is co-chair of the Modeling and Chemistry team of the North American Research Strategy for Tropospheric Ozone (NARSTO).

Armistead G. Russell is a professor of environmental engineering at the Georgia Institute of Technology. His research areas include air-pollution control, aerosol dynamics, atmospheric chemistry, combustion research, control-strategy design, computer modeling, and multiphase fluid dynamics. He received a B.S. from Washington State University and an M.S. and Ph.D. in mechanical engineering at the California Institute of Technology. Dr. Russell was a member of the NRC Committee of Tropospheric Ozone Formation and Measurement. He currently serves as a member of the California Air Resources Board's Reactivity Scientific Advisory Committee.

Christopher L. Saricks is a transportation systems analyst in the Center for Transportation Research in the Energy Systems Division of the Argonne National Laboratory. He received a B.A. from the University of Kansas and an M.Phil. at the University of London. Mr. Saricks' professional activities include mobile-source emission estimates. He currently serves on an NRC review panel for the National Cooperative Highway Research Program.

Appendix B

Letter from Senator Richard G. Lugar

United States Senate

COMMITTEE ON
AGRICULTURE, NUTRITION, AND FORESTRY
WASHINGTON, DC 20510-6000

October 17, 1995

Dr. Stephen Rattien
Executive Director
Commission on Geosciences, Environment and Radiation
National Research Council
2100 Constitution Ave. N.W.
Washington, D.C. 20418

Dear Dr. Rattien:

I understand that the Environmental Protection Agency has requested an evaluation of "whether the existing data is sufficient to show that adding ethanol to RFG on the basis of reactivity would not adversely impact the in-use environmental benefits of the RFG program" and that it has also requested your advice on what additional information would be necessary to allow such a determination to be made. EPA's request is related to a proposal for fuel certification which I discussed at a meeting with Assistant Administrator Mary D. Nichols in November, 1993.

On September 28, 1995 I convened a hearing of the Senate Committee on Agriculture, Nutrition and Forestry on the role of ethanol in the reformulated gasoline program. At that hearing, I discussed EPA's proposal for a National Academy of Sciences (NAS) study with Administrator Carol Browner. Our Committee hearing was called to discuss the restraints which current EPA policies place on refiners who wish to use ethanol blends in reformulated gasoline and the effect of reduced ethanol use on the farm economy and on deficiency payments. Under the RFG program, the EPA judges fuel blends solely by their total mass of emissions of volatile organic compounds ("VOCs"). Since the addition of ethanol to gasoline increases volatility -- and thus increases "evaporative" VOCs -- it is difficult for regular ethanol blends to qualify absent the use of special low RVP gasoline which is more expensive and unavailable in many markets.

However, if one considers the actual ozone forming potential of ethanol blends and not just their mass of emissions, a case can be made that certain ethanol blends may produce reductions in VOCs which are just as great as those produced by qualifying nonethanol blends. Because 10% ethanol blends have greater oxygen content, they may emit fewer exhaust VOCs than non-ethanol

Dr. Stephen Rattien
October 17, 1995
Page Two

blends. And since exhaust VOCs are believed to have a greater
propensity to form ozone than evaporative VOCs, the greater
reduction in exhaust VOCs achieved by certain ethanol RFG blends
may counterbalance their greater mass of evaporative VOCs.
Furthermore, because of the additional oxygen, these blends may
contain less carbon monoxide than non-ethanol blends, further
reducing their tendency to form ozone since carbon monoxide is a
recognized precursor of ozone.

I have proposed that EPA establish a procedure to certify
ethanol blends as equivalent to non-ethanol blends under section
211 (k) (4) (B) of the Clean Air Act, but EPA has so far refused
to do so because it is unsure that there is an appropriate
methodology for making the comparison.

I hope that the NAS Study will have a practical aim; that
is, it will help to determine, in light of the best available
information, the procedures (i.e., the data and analysis) by
which the equivalency of two blends could be determined with a
reasonable degree of certainty. To the extent that additional
information or studies are necessary before such procedures can
be developed or implemented, I also hope that NAS will identify
the additional information and analyses which would be needed and
that it would work with the EPA and with other concerned parties
to ensure that it is provided.

The NAS study is critical to the implementation of the Clean
Air Act in a manner that allows renewable fuels to play an
important role in the reformulated gasoline program. I therefore
join EPA in urging the National Academy of Sciences to undertake
this effort and I urge that it be completed at the earliest
possible date. If you have any questions, please contact me or
Jeff Burnam of my staff at 202-224-7443.

                          Sincerely,

                          Richard G. Lugar
                          Chairman

RGL/jbj

Appendix C

Equation Set for the Complex Model— Phase II RFG[1]

I. *Basic VOC exhaust emissions performance (summer)*

$$\text{VOCE} = \text{VOC(b)} + (\text{VOC(b)} \times Y_{\text{voc}}(t) \div 100) \qquad \text{(C-1)}$$

$$Y_{\text{voc}}(t) = [(w_1 \times N_v) + (w_2 \times H_v) - 1] \times 100 \qquad \text{(C-2)}$$

where

VOCE = exhaust VOC emissions in milligrams per mile

$Y_{\text{voc}}(t)$ = exhaust VOC performance of the target fuel in terms of percentage change from baseline

VOC(b) = baseline (summer) exhaust VOC emissions (= 907.0 mg/mi; see Table 5-6)

N_v = $[\exp v_1(t)] \div [\exp v_1(b)]$

H_v = $[\exp v_2(t)] \div [\exp v_2(b)]$

w_1 = weighting factor for VOC normal-emitter component of fleet (= 0.444)

w_2 = weighting factor for VOC higher-emitter component of fleet (=0.556)

$v_1(t)$ = normal-emitter VOC equation for target fuel, as defined below

---

[1]Adapted from 40 CFR 80.45.

$v_2(t)$ = higher-emitter VOC equation for target fuel, as defined below

$v_1(b)$ = normal-emitter VOC equation, defined below, with base-fuel properties as input

$v_2(b)$ = higher-emitter VOC equation, defined below, with base-fuel properties as input

$\exp(n)$ = the root of Naperian or natural logarithms ($e \approx 2.71828$) raised to the power n.

II. *Consolidated exhaust VOC equations*

For normal emitters:

$$\begin{aligned} v_1 = &(-0.003641 \times OXY) + (0.0005219 \times SUL) + \\ &(0.0289749 \times RVP) + (-0.014470 \times E200) + \\ &(-0.068624 \times E300) + (0.0323712 \times ARO) + \\ &(-0.002858 \times OLE) + (0.0001072 \times E200^2) + \\ &(0.0004087 \times E300^2) + (-0.0003481 \times ARO \times E300) \end{aligned} \quad (C\text{-}3)$$

For higher emitters:

$$\begin{aligned} v_2 = &(-0.003626 \times OXY) + (0.0000540 \times SUL) + \\ &(0.043295 \times RVP) + (-0.013504 \times E200) + \\ &(-0.062327 \times E300) + (0.0282042 \times ARO) + \\ &(-0.002858 \times OLE) + (0.0001060 \times E200^2) + \\ &(0.0004080 \times E300^2) + (-0.0002870 \times ARO \times E300) \end{aligned} \quad (C\text{-}4)$$

where

OXY = oxygen weight percent of fuel

SUL = sulfur content of fuel, in parts per million by weight

RVP = Reid Vapor Pressure of fuel, in pounds per square inch (gauge), measured at 100° F

E200 = 200° F distillation fraction of the fuel, volume percent

E300 = 300° F distillation fraction of the fuel, volume percent

ARO = total aromatics content of fuel, volume percent

OLE = total olefins content of fuel, volume percent.

[NOTE: the value of $Y_{VOC}(t)$ as computed from either of the above equations is modified by linear factoring functions involving deltas (differences between actual and "allowable" values) for E200, E300, and ARO, if any or all of these volume percent values fall outside their allowable ranges.]

III. *Consolidated non-exhaust VOC equations (Phase II)*

For VOC Control Region 1 (south)

$$VOCNE1 = VOCDI1 + VOCHS1 + VOCRL1 + VOCRF1 \qquad (C\text{-}5)$$

$$VOCDI1 = [0.007385 \times RVP^2] - [0.08981 \times RVP] + 0.3158 \quad (C\text{-}6)$$

$$VOCHS1 = [0.006654 \times RVP^2] - [0.08094 \times RVP] + 0.2846 \quad (C\text{-}7)$$

$$VOCRL1 = [0.017768 \times RVP^2] - [0.18746 \times RVP] + 0.6146 \quad (C\text{-}8)$$

$$VOCRF1 = [0.004767 \times RVP] + 0.0.011859 \qquad (C\text{-}9)$$

For VOC Control Region 2 (north)

$$VOCNE2 = VOCDI2 + VOCHS2 + VOCRL2 + VOCRF2 \qquad (C\text{-}10)$$

$$VOCDI2 = [0.004775 \times RVP^2] - [0.05872 \times RVP] + 0.21306 \quad (C\text{-}11)$$

$$VOCHS2 = [0.006078 \times RVP^2] - [0.07474 \times RVP] + 0.27117 \quad (C\text{-}12)$$

$$VOCRL2 = [0.016169 \times RVP^2] - [0.17206 \times RVP] + 0.56724 \quad (C\text{-}13)$$

$$VOCRF2 = [0.004767 \times RVP] + 0.0.011859 \qquad (C\text{-}14)$$

where

VOCNEn = total non-exhaust VOC emissions in control region n, grams per mile

VOCDIn = diurnal[2] VOC emissions in control region n, grams per mile

---

[2]See Chapter 4 for definitions. Measured emissions are apportioned over daily trip distances that are assumed in EPA certification procedures.

VOCHSn = hot soak[2] VOC emissions in control region n, grams per mile

VOCRLn = running loss[2] VOC emissions in control region n, grams per mile

VOCRFn = refueling[2] VOC emissions in control region n, grams per mile.

IV. *Phase II total VOC emissions performance—summer ozone season*

$$\text{VOCSn} = (\text{VOCE} \div 1000) + \text{VOCNEn} \tag{C-15}$$

$$\text{VOCS1\%} = [100\% \times (\text{VOCS1} - 1.4663 \text{ g/mi})] \div 1.4663 \text{ g/mi} \tag{C-16}$$

$$\text{VOCS2\%} = [100\% \times (\text{VOCS2} - 1.3991 \text{ g/mi})] \div 1.3991 \text{ g/mi} \tag{C-17}$$

where

VOCSn = total summer VOC emissions in control region n, grams per mile; VOCE, VOCNEn as defined above

VOCS1% = total summer VOC emissions *performance* of target fuel for VOC control Region 1 (south), in percentage terms relative to baseline level

VOCS2% = total summer VOC emissions *performance* of target fuel for VOC control Region 2 (north), in percentage terms relative to baseline level.

V. *Summer NO_x emissions performance*

$$NO_x = NO_x(b) + [NO_x(b) \times Y(t) \div 100) \tag{C-18}$$

$$Y_{NOX}(t) = [(z_1 \times N_n) + (z_2 \times H_n) - 1] \times 100 \tag{C-19}$$

where

NO_x = exhaust NO_x emissions in milligrams per mile

$Y_{NOX}(t)$ = NO_x performance of the target fuel in terms of percentage change from baseline

$NO_x(b)$ = baseline NO_x emissions (=1340 mg/mi, see Table 5-6)

N_n = [exp $n_1(t)$] ÷ [exp $n_1(b)$]

H_v = [exp $n_2(t)$] ÷ [exp $n_2(b)$]

z_1 = weighting factor for NO_x normal-emitter component of fleet (=0.738)

z_2 = weighting factor for NO_x higher-emitter component of fleet (=0.262)

$n_1(t)$ = normal-emitter NO_x equation for target fuel, as defined below

$n_2(t)$ = higher-emitter NO_x equation for target fuel, as defined below

$n_1(b)$ = normal-emitter NO_x equation, defined below, with base-fuel properties as input

$n_2(b)$ = higher-emitter NO_x equation, defined below, with base-fuel properties as input.

VI. *Consolidated NO_x equations*

For normal emitters:

$$
\begin{aligned}
n_1 = \;& (0.0018571 \times OXY) + (0.0006921 \times SUL) + \\
& (0.0090744 \times RVP) + (0.0009310 \times E200) + \\
& (0.0008460 \times E300) + (0.0083632 \times ARO) - \\
& (0.002774 \times OLE) - (0.000000663 \times SUL^2) - \\
& (0.000119 \times ARO^2) + (0.0003665 \times OLE^2)
\end{aligned}
\qquad \text{(C-20)}
$$

For higher emitters:

$$
\begin{aligned}
n_2 = \;& (-0.00913 \times OXY) + (0.000252 \times SUL) - \\
& (0.01397 \times RVP) + (0.000931 \times E200) - \\
& (0.00401 \times E300) + (0.007097 \times ARO) - \\
& (0.00276 \times OLE) + (0.0003665 \times OLE^2) - \\
& (0.00007995 \times ARO^2)
\end{aligned}
\qquad \text{(C-21)}
$$

[NOTE: the value of $Y_{NOX}(t)$ as computed from either of the above equations is modified by linear factoring functions involving deltas (differences between actual and "allowable" values) for SUL, OLE, and ARO, if any or all of these volume percent values fall outside their allowable ranges.]

VII. *Summer toxics emissions performance, Phase II*

TOXICSn   = EXHBZ + FORM + ACET + BUTA + POM + NEBZn

$$\text{(C-22)}$$

TOXICS1% = [100% x (TOXICS1 - 86.34 mg/mi)] ÷ 86.34 mg/mi

$$\text{(C-23)}$$

TOXICS2% = [100% x (TOXICS2 - 85.61 mg/mi)] ÷ 85.61 mg/mi

$$\text{(C-24)}$$

where

TOXICSn	= summer toxics performance, VOC Control Region n, milligrams per mile
TOXICSn%	= TOXICS performance of the target fuel in VOC Control Region n, in terms of percentage change from baseline
EXHBZ	= exhaust emissions of benzene as computed below, milligrams per mile
FORM	= exhaust emissions of formaldehyde as computed below, milligrams per mile
ACET	= exhaust emissions of acetaldehyde as computed below, milligrams per mile
BUTA	= exhaust emissions of 1,3-butadiene as computed below, milligrams per mile
POM	= exhaust emissions of polycyclic organic matter as computed below, milligrams per mile
NEBZn	= non-exhaust emissions of benzene, VOC Control Region n, as computed below, milligrams per mile.

VIII. *Emissions equations for individual ozone-season toxics—*
    (1) benzene

$$\text{EXHBZ} = \text{BENZ(b)} + (\text{BENZ(b)} \times Y_{BEN}(t) \div 100) \qquad \text{(C-25)}$$

$$Y_{BEN}(t) = [(w_1 \times N_b) + (w_2 \times H_b) - 1] \times 100 \qquad \text{(C-26)}$$

where

$Y_{BEN}(t)$ = exhaust benzene performance of the target fuel in terms of percentage change from baseline

$BENZ(b)$ = baseline (summer) exhaust benzene emissions (=53.54 mg/mi, from Table 5-6)

N_b = $[\exp b_1(t)] \div [\exp b_1(b)]$

H_b = $[\exp b_2(t)] \div [\exp b_2(b)]$

w_1 = weighting factor for toxics normal-emitter component of fleet (=0.444)

w_2 = weighting factor for toxics higher-emitter component of fleet (=0.556)

$b_1(t)$ = normal-emitter benzene equation for target fuel, as defined below

$b_2(t)$ = higher-emitter benzene equation for target fuel, as defined below

$b_1(b)$ = normal-emitter benzene equation, defined below, with base-fuel properties as input

$b_2(b)$ = higher-emitter benzene equation, defined below, with base-fuel properties as input.

IX. *Consolidated benzene equations*

For normal emitters:

$$b_1 = (0.0006197 \times SUL) - (0.003376 \times E200) + (0.0265500 \times ARO) + (0.2223900 \times BEN) \qquad \text{(C-27)}$$

For higher emitters:

$$n_2 = (-0.096047 \times OXY) + (0.0003370 \times SUL) - (0.0112510 \times E300) + (0.0118820 \times ARO) + (0.2223180 \times BEN) \qquad \text{(C-28)}$$

where BEN = benzene content of target fuel, volume percent, and all other terms are as defined above.

X. *Emissions equations for individual ozone-season toxics—*
    (2) formaldehyde

$$\text{FORM} = \text{FORM}(b) + (\text{FORM}(b) \times Y_{\text{FORM}}(t) \div 100) \qquad \text{(C-29)}$$

$$Y_{\text{FORM}}(t) = [(w_1 \times N_f) + (w_2 \times H_f) - 1] \times 100 \qquad \text{(C-30)}$$

where

$Y_{\text{FORM}}(t)$ = exhaust formaldehyde performance of the target fuel in terms of percentage change from baseline

$\text{FORM}(b)$ = baseline (summer) exhaust formaldehyde emissions (=9.70 mg/mi, see Table 5-6)

N_f = $[\exp f_1(t)] \div [\exp f_1(b)]$

H_f = $[\exp f_2(t)] \div [\exp f_2(b)]$

$f_1(t)$ = normal-emitter formaldehyde equation for target fuel, as defined below

$f_2(t)$ = higher-emitter formaldehyde equation for target fuel, as defined below

$f_1(b)$ = normal-emitter formaldehyde equation below, with base-fuel properties as input

$f_2(b)$ = higher-emitter formaldehyde equation below, with base-fuel properties as input.

XI. *Consolidated formaldehyde equations*

For normal emitters:

$$f_1 = (-0.010226 \times \text{E}300) - (0.007166 \times \text{ARO}) + (0.0462131 \times \text{MTB}) \qquad \text{(C-31)}$$

For higher emitters:

$$f_2 = (-0.010226 \times \text{E}300) - (0.007166 \times \text{ARO}) - (0.031352 \times \text{OLE}) + (0.0462131 \times \text{MTB}) \qquad \text{(C-32)}$$

where MTB = methyl tertiary-butyl ether content of target fuel, weight percent oxygen, and all other terms are as defined above.

XII. *Emissions equations for individual ozone-season toxics—*
(3) *acetaldehyde*

$$ACET = ACET(b) + (ACET\ (b) \times Y_{ACET}(t) \div 100) \qquad (C\text{-}33)$$

$$Y_{ACRT}(t) = [(w_1 \times N_a) + (w_2 \times H_a) - 1] \times 100 \qquad (C\text{-}34)$$

where

$Y_{ACET}(t)$ = Exhaust acetaldehyde performance of the target fuel in terms of percentage change from baseline

$ACET(b)$ = baseline (summer) exhaust acetaldehyde emissions (=4.44 mg/mi, see Table 5-6)

N_a = [exp $a_1(t)$] ÷ [exp $a_1(b)$]

H_a = [exp $a_2(t)$] ÷ [exp $a_2(b)$]

$a_1(t)$ = normal-emitter acetaldehyde equation for target fuel, as defined below

$a_2(t)$ = higher-emitter acetaldehyde equation for target fuel, as defined below

$a_1(b)$ = normal-emitter acetaldehyde equation below, with base-fuel properties as input

$a_2(b)$ = higher-emitter acetaldehyde equation below, with base-fuel properties as input.

XIII. *Consolidated acetaldehyde equations*

For normal emitters:

$$
\begin{aligned}
a_1 ={}& (0.0002631 \times SUL) + (0.0397860 \times RVP) - \\
& (0.012172 \times E300) - (0.005525 \times ARO) - \\
& (0.009594 \times MTB) + (0.3165800 \times ETB) + \\
& (0.2492500 \times ETH) \qquad\qquad\qquad (C\text{-}35)
\end{aligned}
$$

For higher emitters:

$$
\begin{aligned}
a_2 ={}& (0.0002627 \times SUL) - (0.012157 \times E300) - \\
& (0.005548 \times ARO) - (0.055980 \times MTB) + \\
& (0.3164665 \times ETB) + (0.2493259 \times ETH) \qquad (C\text{-}36)
\end{aligned}
$$

where ETB = ethyl tertiary-butyl ether content of target fuel, weight percent oxygen and ETH = ethanol content of target fuel, weight percent oxygen, and all other terms are as defined above.

XIV. *Emissions equations for individual ozone-season toxics—*
     (4) 1,3-butadiene

$$\text{BUTA} = \text{BUTA(b)} + (\text{BUTA (b)} \times Y_{\text{BUTA}}(t) \div 100) \qquad \text{(C-37)}$$

$$Y_{\text{BUTA}}(t) = [(w_1 \times N_d) + (w_2 \times H_d) - 1] \times 100 \qquad \text{(C-38)}$$

where

$Y_{\text{BUTA}}(t)$ = Exhaust 1,3-butadiene performance of the target fuel in terms of percentage change from baseline

BUTA(b) = Baseline (summer) exhaust 1,3-butadiene emissions (=9.38 mg/mi, see Table 5-6)

N_d = $[\exp d_1(t)] \div [\exp d_1(b)]$

H_d = $[\exp d_2(t)] \div [\exp d_2(b)]$

$d_1(t)$ = normal-emitter 1,3-butadiene equation for target fuel, as defined below

$d_2(t)$ = higher-emitter 1,3-butadiene equation for target fuel, as defined below

$d_1(b)$ = normal-emitter 1,3-butadiene equation below, with base-fuel properties as input

$d_2(b)$ = higher-emitter 1,3-butadiene equation below, with base-fuel properties as input.

XV. *Consolidated 1,3-butadiene equations*

For normal emitters:

$$d_1 = (0.0001552 \times \text{SUL}) - (0.007253 \times \text{E200}) -$$
$$(0.014866 \times \text{E300}) - (0.004005 \times \text{ARO}) +$$
$$(0.028235 \times \text{OLE}) \qquad \text{(C-39)}$$

For higher emitters:

d_2 = $-0.060771 \times$ OXY) $- (0.007311 \times$ E200) $-$
              $(0.008058 \times$ E300) $- (0.004005 \times$ ARO) $+$
              $(0.0436960 \times$ OLE)   (C-40)

where OXY = oxygen content of target fuel, weight percent, and all
other terms are as defined above.

XVI. *Polycyclic organic matter, mass emissions (milligrams per mile)*

POM   = $0.003355 \times$ VOCE   (C-41)

Terms are as defined above.

XVII. *Non-exhaust benzene emissions (milligrams per mile)*

NEBZn   = DIBZn + HSBZn + RLBZn + RFBZn   (C-42)

where terms are defined as under Part III above, but "BZ" refers only
to the benzene component of evaporative emissions.

For VOC Control Region 1:

DIBZ1   = $10 \times$ BEN $\times$ VOCDI1 $\times$ $[(-0.0290 \times$ MTB) $-$
              $(0.080274 \times$ RVP) $+ 1.3758]$ (C-43)

HSBZ1   = $10 \times$ BEN $\times$ VOCHS1 $\times$ $[(-0.0342 \times$ MTB) $-$
              $(0.080274 \times$ RVP) $+ 1.4448]$ (C-44)

RLBZ1   = $10 \times$ BEN $\times$ VOCRL1 $\times$ $[(-0.0342 \times$ MTB) $-$
              $(0.080274 \times$ RVP) $+ 1.4448]$ (C-45)

RFBZ1   = $10 \times$ BEN $\times$ VOCRF1 $\times$ $[(-0.0296 \times$ MTB) $-$
              $(0.081507 \times$ RVP) $+ 1.3972]$ (C-46)

For VOC Control Region 2:

DIBZ2   = $10 \times$ BEN $\times$ VOCDI2 $\times$ $[(-0.0290 \times$ MTB) $-$

$$(0.080274 \times RVP) + 1.3758] \qquad \text{(C-47)}$$

$$HSBZ2 = 10 \times BEN \times VOCHS2 \times [(-0.0342 \times MTB) -$$
$$(0.080274 \times RVP) + 1.4448] \qquad \text{(C-48)}$$

$$RLBZ2 = 10 \times BEN \times VOCRL2 \times [(-0.0342 \times MTB) -$$
$$(0.080274 \times RVP) + 1.4448] \qquad \text{(C-49)}$$

$$RFBZ2 = 10 \times BEN \times VOCRF2 \times [(-0.0296 \times MTB) -$$
$$(0.081507 \times RVP) + 1.3972] \qquad \text{(C-50)}$$

All terms are as defined above.

[NOTE: For purposes of comparing weight percent vs. volume percent of oxygen, approximate conversion values are the following: for MTBE as oxygenate, $W_m = V_m \times 0.18$, and for ethanol as oxygenate, $W_e = V_e \times 0.35$, where W is weight percent and V is volume percent.]

Appendix D

Data on Emissions from Light-Duty Motor Vehicles Using Fuels Selected from the Auto/Oil Air Quality Improvement Research Program and the California Ethanol Testing Program

TABLE D-1 Exhaust Emissions from Fuels F, S, U, T, N2 and MM

Fuel Type	No.	90 MIR[a] (g O₃/mi) mean ± SDM	91 MIR[b] (g O₃/mi) mean ± SDM	97 MIR[c] (g O₃/mi) mean ± SDM	min	max	NOₓ (g/mi) mean ± SDM	min	max	VOC (g/mi) mean ± SDM	min	max
No Oxygenate												
F	19	0.480 ± 0.070	0.663 ± 0.096	0.766 ± 0.117	0.34	2.30	0.530 ± 0.065	0.10	1.05	0.201 ± 0.028	0.10	0.57
S	9	0.496 ± 0.105	0.687 ± 0.146	0.820 ± 0.182	0.43	2.24	0.566 ± 0.083	0.13	0.94	0.218 ± 0.047	0.10	0.58
Ethanol Blend												
U	11	0.476 ± 0.097	0.658 ± 0.133	0.765 ± 0.164	0.31	2.20	0.574 ± 0.087	0.13	1.06	0.205 ± 0.040	0.10	0.56
T	10	0.465 ± 0.096	0.644 ± 0.133	0.757 ± 0.165	0.40	2.20	0.572 ± 0.095	0.14	1.11	0.203 ± 0.040	0.11	0.55
MTBE Blend												
N2	9	0.497 ± 0.114	0.681 ± 0.156	0.803 ± 0.188	0.39	2.24	0.619 ± 0.095	0.11	1.05	0.206 ± 0.042	0.11	0.53
MM	12	0.439 ± 0.084	0.601 ± 0.115	0.703 ± 0.139	0.38	2.11	0.583 ± 0.073	0.14	0.85	0.183 ± 0.031	0.11	0.50

Fuel Type	No.	CO (g/mi) mean ± SDM	min	max	97 MIR Specific Reactivity (g O₃/g VOC) mean ± SDM	min	max
No Oxygenate							
F	19	2.76 ± 0.50	0.95	7.92	2.96 ± 0.12	1.68	3.79
S	9	2.57 ± 0.57	1.08	5.68	3.02 ± 0.15	1.99	3.74
Ethanol Blend							
U	11	2.79 ± 0.72	0.81	8.29	2.85 ± 0.13	1.87	3.66
T	10	2.52 ± 0.60	0.81	6.42	2.93 ± 0.15	1.85	3.62
MTBE Blend							
N2	9	2.82 ± 0.73	0.95	6.86	3.00 ± 0.16	2.00	3.75
MM	12	2.28 ± 0.54	0.74	6.30	3.03 ± 0.16	1.98	4.00

[a] Uses the SAPRC1990 chemical mechanism.
[b] Uses the SAPRC1991 chemical mechanism.
[c] Uses the SAPRC1997 chemical mechanism.
Abbreviations: No., number of samples; SDM, standard deviation of the mean; min, minimum; max, maximum.

TABLE D-2 Diurnal Emissions for Fuels F, S, U, T, N2 and MM

Fuel Type	No.	97 MIR[a] (g O$_3$/test)			VOC (g/test)		
		mean ± SDM	min	max	mean ± SDM	min	max
No Oxygenate							
F	26	0.782 ± 0.092	0.19	1.89	0.311 ± 0.035	0.12	0.88
S	9	0.556 ± 0.116	0.19	1.41	0.199 ± 0.034	0.07	0.37
Ethanol Blend							
U	9	0.801 ± 0.120	0.44	1.56	0.305 ± 0.040	0.20	0.55
T	9	0.699 ± 0.113	0.31	1.25	0.256 ± 0.040	0.15	0.51
MTBE Blend							
N2	9	0.680 ± 0.196	0.19	1.75	0.221 ± 0.054	0.06	0.55
MM	9	0.737 ± 0.124	0.29	1.49	0.228 ± 0.035	0.10	0.40

Fuel Type	No.	97 MIR Specific Reactivity (g O$_3$/g VOC)		
		mean ± SDM	min	max
No Oxygenate				
F	26	2.55 ± 0.14	1.40	4.05
S	9	2.76 ± 0.21	1.99	3.86
Ethanol Blend				
U	9	2.60 ± 0.14	1.76	3.16
T	9	2.71 ± 0.14	2.00	3.34
MTBE Blend				
N2	9	3.06 ± 0.37	1.52	5.63
MM	9	3.23 ± 0.22	2.19	4.12

[a]Uses the SAPRC1997 chemical mechanism.

TABLE D-3   Hot-Soak Emissions for Fuels F, S, U, T, N2 and MM

Fuel Type	No.	97 MIR[a] (g O_3/test)			VOC (g/test)		
		mean ± SDM	min	max	mean ± SDM	min	max
No Oxygenate							
F	26	1.10 ± 0.21	0.16	4.37	0.306 ± 0.054	0.07	1.11
S	9	0.92 ± 0.29	0.20	3.12	0.257 ± 0.075	0.06	0.84
Ethanol Blend							
U	9	1.36 ± 0.27	0.37	3.17	0.422 ± 0.082	0.14	0.96
T	9	1.26 ± 0.30	0.36	3.55	0.385 ± 0.091	0.12	1.08
MTBE Blend							
N2	9	0.97 ± 0.31	0.18	3.30	0.282 ± 0.085	0.07	0.93
MM	9	1.27 ± 0.32	0.21	3.55	0.339 ± 0.084	0.07	0.91

Fuel Type	97 MIR Specific Reactivity (g O_3/g VOC)		
	mean ± SDM	min	max
No Oxygenate			
F	3.43 ± 0.09	2.48	4.40
S	3.49 ± 0.08	3.11	3.84
Ethanol Blend			
U	3.17 ± 0.10	2.56	3.58
T	3.28 ± 0.10	2.76	3.86
MTBE Blend			
N2	3.31 ± 0.12	2.73	3.93
MM	3.73 ± 0.12	3.01	4.18

[a]Uses the SAPRC1997 chemical mechanism.

TABLE D-4 Exhaust Emissions for Current, Federal Tier 1, and Advanced Technology Vehicles Using Fuels C1 and C2

Fleet	Fuel C1 (no oxygenate)			Fuel C2 (MTBE)		
	Mean ± SDM	min	max	Mean ± SDM	min	max
Current	(No. = 9)			(No. = 17)		
1991 MIR (g O_3/mi)	0.808 ± 0.193	0.37	2.28	0.783 ± 0.137	0.32	2.32
1997 MIR (g O_3/mi)	0.969 ± 0.225	0.45	2.68	0.964 ± 0.165	0.42	2.69
NO_x (g/mi)	0.466 ± 0.080	0.13	0.84	0.494 ± 0.067	0.09	0.99
VOC (g/mi)	0.239 ± 0.050	0.10	0.61	0.241 ± 0.038	0.10	0.65
97 MIR specific reactivity (g O_3/g VOC)	3.38 ± 0.17	2.23	4.01	3.42 ± 0.13	2.31	4.09
CO (g/mi)	2.46 ± 0.65	0.77	6.38	2.21 ± 0.40	0.86	6.20
HCHO (mg/mi)	4.9 ± 1.2	1.7	14.1	4.9 ± 0.8	1.3	13.6
CH_3CHO (mg/mi)	1.6 ± 0.4	0.7	4.6	1.6 ± 0.3	0.7	4.7
Butadiene (mg/mi)	1.02 ± 0.23	0.4	2.7	0.97 ± 0.18	0.4	2.9
Benzene (mg/mi)	6.6 ± 1.8	2.8	21.0	7.3 ± 1.6	2.6	25.9
Federal Tier 1	(No. = 10)			(No. = 10)		
1991 MIR (g O_3/mi)	0.423 ± 0.025	0.34	0.49	0.416 ± 0.022	0.27	0.51
1997 MIR (g O_3/mi)	0.518 ± 0.027	0.42	0.59	0.505 ± 0.026	0.38	0.69
NO_x (g/mi)	0.284 ± 0.052	0.16	0.51	0.316 ± 0.049	0.13	0.55
VOC (g/mi)	0.121 ± 0.005	0.10	0.14	0.118 ± 0.006	0.09	0.16
97 MIR specific reactivity (g O_3/g VOC)	3.51 ± 0.14	2.93	3.83	3.52 ± 0.11	2.97	4.04
CO (g/mi)	1.35 ± 0.12	0.99	1.94	1.34 ± 0.11	0.78	2.19
HCHO (mg/mi)	1.8 ± 0.2	0.9	2.3	2.1 ± 0.1	1.2	2.6
CH_3CHO (mg/mi)	0.66 ± 0.08	0.3	0.9	0.63 ± 0.06	0.3	1.0
Butadiene (mg/mi)	0.59 ± 0.08	0.3	0.9	0.55 ± 0.06	0.3	0.8

TABLE D-4 Exhaust Emissions for Current, Federal Tier 1, and Advanced Technology Vehicles Using Fuels C1 and C2 (Continued)

Fleet	Fuel C1 (no oxygenate)			Fuel C2 (MTBE)		
	Mean ± SDM	min	max	Mean ± SDM	min	max
Federal Tier 1 (Continued)	(No. = 10)			(No. = 10)		
Benzene (mg/mi)	3.3 ± 0.2	2.5	3.8	3.4 ± 0.2	2.5	4.2
Advanced Technology	(No. = 12)			(No. = 12)		
1991 MIR (g O_3/mi)	0.298 ± 0.028	0.19	0.39	0.280 ± 0.022	0.14	0.40
1997 MIR (g O_3/mi)	0.371 ± 0.025	0.28	0.45	0.349 ± 0.026	0.19	0.51
NO_x (g/mi)	0.124 ± 0.029	0.04	0.25	0.122 ± 0.019	0.03	0.26
VOC (g/mi)	0.088 ± 0.005	0.07	0.11	0.083 ± 0.006	0.06	0.12
97 MIR Specific Reactivity (g O_3/g VOC)	3.56 ± 0.19	2.79	4.18	3.53 ± 0.11	2.70	4.04
CO (g/mi)	0.90 ± 0.12	0.58	1.39	0.84 ± 0.07	0.48	1.26
HCHO (mg/mi)	1.4 ± 0.4	0.5	3.5	1.6 ± 0.2	0.7	3.4
CH_3CHO (mg/mi)	0.44 ± 0.12	0.2	1.0	0.48 ± 0.09	0.1	1.2
Butadiene (mg/mi)	0.44 ± 0.07	0.3	0.8	0.42 ± 0.06	0.2	0.9
Benzene (mg/mi)	2.8 ± 0.2	2.2	3.8	2.7 ± 0.1	2.0	3.3

TABLE D-5 Hot-Soak Emissions for Current Vehicles, Federal Tier 1 Vehicles, and Advanced-Technology Vehicles Using Fuels C1 and C2

Fleet	Fuel C1 (no oxygenate)			Fuel C2 (MTBE)		
	Mean ± SDM	min	max	Mean ± SDM	min	max
Current	(No. = 8)			(No. = 17)		
1991 MIR (g O₃/test)	0.360 ± 0.049	0.21	0.56	0.417 ± 0.091	0.03	1.80
1997 MIR (g O₃/test)	0.520 ± 0.085	0.14	0.94	0.575 ± 0.102	0.07	2.11
VOC (g/test)	0.13 ± 0.02	0.05	0.24	0.16 ± 0.03	0.02	0.52
97 MIR specific reactivity (g O₃/g VOC)	3.90 ± 0.20	2.62	4.70	3.60 ± 0.15	1.67	4.44
Benzene (mg/test)	3.7 ± 0.6	1.7	6.8	4.0 ± 0.9	0.3	17.6
Federal Tier 1	(No. = 6)			(No. = 11)		
1991 MIR (g O₃/test)	0.566 ± 0.172	0.28	1.49	0.659 ± 0.150	0.15	1.68
1997 MIR (g O₃/test)	0.748 ± 0.241	0.37	2.05	0.805 ± 0.198	0.19	2.13
VOC (g/test)	0.18 ± 0.06	0.08	0.50	0.21 ± 0.05	0.06	0.57
97 MIR specific reactivity (g O₃/g VOC)	4.07 ± 0.15	3.54	4.49	3.79 ± 0.11	3.23	4.41
Benzene (mg/test)	5.9 ± 2.3	2.0	18.1	6.1 ± 1.6	1.1	18.5
Advanced Technology	(No. = 1)			(No. = 3)		
1991 MIR (g O₃/test)	2.20			2.11 ± 0.41	1.47	3.11
1997 MIR (g O₃/test)	2.96			2.62 ± 0.46	2.04	3.75
VOC (g/test)	0.94			0.62 ± 0.11	0.48	0.89
g O₃/g VOC	3.16			4.27 ± 0.12	4.24	4.29
Benzene (mg/test)	17.8			11.6 ± 1.8	7.5	15.1

TABLE D-6 CARB Emission Testing Results of the FTP Test

Vehicle No.	Fuel 63 Total Reactivity O_3 (mg/mi)	Specific Reactivity	Log Total Reactivity	Log Specific Reactivity	Fuel 64 Total Reactivity O_3 (mg/mi)	Specific Reactivity	Log Total Reactivity	Log Specific Reactivity
2	1033	3.84			1507	3.85	3.18	0.59
	1198	3.78			1812	3.51	3.26	0.55
	1160	3.81			*Mean* 1660	3.68	3.22	0.57
	Mean 1130	3.81	3.05	0.58				
3	249	3.66			329	3.56		
	215	3.45			360	3.32		
	Mean 232	3.55	2.37	0.55	*Mean* 344	3.44	2.54	0.54
4	644	3.46			906	3.30		
	629	3.56			590	3.55		
	Mean 637	3.51	2.80	0.55	722	3.41		
					Mean 739	3.42	2.87	0.53
5	895	3.16			816	3.20		
	870	3.33			1487	3.20		
	Mean 882	3.25	2.95	0.51	1225	2.95		
					Mean 1176	3.12	3.07	0.49
6	744	3.29			547	3.31		
	572	3.54			663	3.24		
	610	3.78			518	3.56		
	Mean 642	3.54	2.81	0.55	*Mean* 576	3.37	2.76	0.53
7	274	3.38			250	3.44		
	314	3.28			281	3.32		
	Mean 294	3.33	2.47	0.52	*Mean* 265	3.38	2.42	0.53
8	1199	3.37			1088	3.50		

8	1056	3.48			1044	3.48		
9	*Mean* 1128	3.42	*3.05*	*0.53*	1066	3.49	*3.03*	*0.54*
	664	3.21			449	3.44		
9	492	3.42			476	3.37		
10	*Mean* 578	3.31	*2.76*	*0.52*	462	3.41	*2.67*	*0.53*
	584	3.56			647	3.69		
	577	3.41			624	3.84		
	569	3.76						
10	*Mean* 577	3.58	*2.76*	*0.55*	635	3.76	*2.80*	*0.58*
11	552	3.81			491	3.69		
	469	3.80			612	3.46		
11	*Mean* 511	3.80	*2.71*	*0.58*	552	3.58	*2.74*	*0.55*
13	521	3.43			615	3.71		
	544	3.55			592	3.73		
	595	3.64						
13	*Mean* 553	3.54	*2.74*	*0.55*	604	3.72	*2.78*	*0.57*
14	941	3.42			516	3.56		
	706	4.32			465	3.65		
	549	3.56						
14	*Mean* 732	3.76	*2.86*	*0.58*	490	3.61	*2.69*	*0.56*
	Mean 658	3.53	*2.78*	*0.548*	714	3.50	*2.80*	*0.543*
	sd 280	0.189	*0.205*	*0.0232*	399	0.181	*0.224*	*0.0228*
	sd 81.7	0.0544	*0.0591*	*0.00668*	115	0.0523	*0.0647*	*0.000658*
	mean							
t test	F-64-F-63	F-64-F-63	*F-64-F-63*	*F-64-F-63*				
p value	0.35	0.40	*0.50*	*0.40*				

TABLE D-7 CARB Emission-Testing Results of the Rep05 Cell

Vehicle No.		Fuel 63				Fuel 64			
		Total Reactivity O₃ (mg/mi)	Specific Reactivity	Log Total Reactivity	Log Specific Reactivity	Total Reactivity O₃ (mg/mi)	Specific Reactivity	Log Total Reactivity	Log Specific Reactivity
2		180.44	3.83	2.26	0.58	151.31	3.88	2.18	0.59
		130.64	3.76	2.12	0.57	102.47	3.73	2.01	0.57
	Mean	155.54	3.79	*2.19*	*0.58*	126.89	3.81	*2.10*	*0.58*
3		166.99	4.17			101.98	4.17		
		118.38	4.01			105.49	4.10		
	Mean	142.68	4.09	*2.15*	*0.61*	103.74	4.13	*2.02*	*0.62*
4		340.06	3.88			380.06	3.93		
		366.25	4.00			320.68	3.89		
	Mean	353.16	3.94	*2.55*	*0.60*	354.32	3.91		
5		908.54	3.30			351.68	3.91	*2.55*	*0.59*
		845.40	3.24			756.94	3.09		
		1048.35	3.07			926.93	2.89		
	Mean	934.10	3.20	*2.97*	*0.51*	841.93	2.99	*2.93*	*0.48*
6		158.42	4.08			143.91	4.18		
		128.46	4.06			125.13	4.02		
		144.61	4.12			99.74	3.83		
	Mean	143.83	4.09	*2.16*	*0.61*	122.93	4.01	*2.09*	*0.60*
7		116.42	3.47			73.04	3.47		
		124.17	1.99			121.12	3.48		
	Mean	120.29	2.73	*2.08*	*0.44*	97.08	3.48	*1.99*	*0.54*
8		273.45	3.61			292.15	3.57		
		298.20	3.52			407.08	3.73		

	Mean		t test	p value	Mean		t test	p value
9	285.83	3.57	2.46	0.55	349.61	3.65	2.54	0.56
	85.89	2.77			105.84	3.02		
	80.83	2.93			73.51	3.10		
10	83.36	2.85	1.92	0.45	89.67	3.06	1.95	0.49
	214.00	3.13			103.53	2.81		
	211.34	3.00			148.80	3.20		
	139.07	2.76						
11	188.13	2.96	2.27	0.47	126.17	3.01	2.10	0.48
	265.93	3.97			245.14	3.96		
	269.96	3.95			285.10	4.11		
13	267.94	3.96	2.43	0.60	265.12	4.04	2.42	0.61
	27.52	3.05			43.75	3.37		
	24.70	3.08			19.26	2.73		
	39.54	3.44						
14	30.59	3.19	1.49	0.50	31.50	3.05	1.50	0.48
	176.44	3.33			230.01	3.44		
	194.07	3.45			187.81	3.38		
	106.58	2.96						
Mean	159.03	3.25	2.20	0.51	208.91	3.41	2.32	0.53
Mean	239	3.47	2.23	0.539	226	3.26	2.21	0.48
sd	236	0.498	0.346	0.0615	220	0.932	0.37	0.22
sdm	68.3	0.144	0.100	0.0178	63.4	0.269	0.106	0.065
t test	F-64-F-63	F-64-F-63	F-64-F-63	F-64-F-63	F-64-F-63			
p value	0.34	0.51	0.26	0.43				

TABLE D-8 CARB Emission-Testing Results of the 1-Hour Hot Soak (SHED)

Vehicle No.	Fuel 63 Total Reactivity O₃ (mg/mi)	Specific Reactivity	Log Total Reactivity	Log Specific Reactivity	Fuel 64 Total Reactivity O₃ (mg/mi)	Specific Reactivity	Log Total Reactivity	Log Specific Reactivity
7	687.39	2.87			793.28	2.91		
7	774.40	2.89			1091.27	2.65		
Mean	*730.89*	*2.88*	*2.86*	*0.46*	*942.28*	*2.78*	*2.97*	*0.44*
8	112.78	2.08			127.00	2.17		
8	103.09	2.57			241.20	2.25		
Mean	*107.94*	*2.32*	*2.03*	*0.37*	*184.10*	*2.21*	*2.27*	*0.34*
9	204.81	2.65			546.12	2.23		
9	220.84	2.65			496.77	2.39		
Mean	*212.83*	*2.65*	*2.33*	*0.42*	*521.44*	*2.31*	*2.72*	*0.36*
10	873.68	2.90			1762.09	2.75		
10	988.00	2.82			1679.19	2.61		
10	964.17	2.86						
Mean	*941.95*	*2.86*	*2.97*	*0.46*	*1720.64*	*2.68*	*3.24*	*0.43*
11	138.64	2.66			165.89	2.16		
11	111.06	2.32			223.55	2.24		
Mean	*124.85*	*2.49*	*2.10*	*0.40*	*194.72*	*2.20*	*2.29*	*0.34*
13	220.96	2.87			384.84	2.26		
13	133.76	2.58			404.01	2.22		
13	142.61	2.56						
Mean	*165.78*	*2.67*	*2.22*	*0.43*	*394.43*	*2.24*	*2.60*	*0.35*
Mean	*381*	*2.65*	*2.42*	*0.421*	*660*	*2.40*	*2.68*	*0.379*
sd	*361*	*0.213*	*0.402*	*0.0356*	*589*	*0.259*	*0.382*	*0.0454*
sdm	*147*	*0.087*	*0.164*	*0.0145*	*241*	*0.106*	*0.156*	*0.0186*
t test	F-64-F-63	F-64-F-63	F-64-F-63	F-64-F-63				
p value	0.048	0.0063	0.0020	0.0075				

Note: SHED = sealed-housing-for-evaporative-determination facility.

TABLE D-9 CARB Emission-Testing Results of the 0- to 24-Hour Diurnal (SHED)

Vehicle No.	Fuel 63				Fuel 64			
	Total Reactivity O₃ (mg/mi)	Specific Reactivity	Log Total Reactivity	Log Specific Reactivity	Total Reactivity O₃ (mg/mi)	Specific Reactivity	Log Total Reactivity	Log Specific Reactivity
7	21223.40	2.15			21871.28	2.00		
	18659.70	2.13			24969.49	2.10		
Mean	19941.55	2.14	4.30	0.33	23420.38	2.05	4.37	0.31
8	4282.02	1.41			11250.40	1.18		
	4893.90	1.33			8844.45	1.30		
Mean	4587.96	1.37	3.66	0.14	10047.42	1.24	4.00	0.09
9	2650.33	1.47			4531.86	1.83		
	2643.23	1.52			3869.91	2.07		
Mean	2646.78	1.50	3.42	0.18	4200.89	1.95	3.62	0.29
10	13680.52	2.15			33992.48	1.90		
	14127.95	2.11			28313.19	1.71		
	19181.15	1.74						
Mean	15663.20	2.00	4.19	0.30	31152.83	1.81	4.49	0.26
11	2182.50	1.84			2921.51	1.92		
	2003.68	2.01			3071.85	2.03		
Mean	2093.09	1.93	3.32	0.28	2996.68	1.97	3.48	0.30
13	2622.41	1.15			2872.04	1.50		
	1583.05	1.29			3334.12	1.40		
	1873.49	1.25						
Mean	2026.32	1.23	3.31	0.09	3103.08	1.45	3.49	0.16
Mean	7830	1.69	3.70	0.219	12490	1.74	3.91	0.235
sd	7900	0.377	0.443	0.0992	12000	0.327	0.448	0.0880
sdm	3230	0.154	0.181	0.0405	4900	0.134	0.183	0.0359
t test	F-64-F-63	F-64-F-63	F-64-F-63	F-64-F-63				
p value	0.10	0.63	0.0035	0.59				

TABLE D-10 CARB Emission Testing Results of the 24- to 48-Hour Diurnal (SHED)

Vehicle No.		Fuel 63 Total Reactivity O_3 (mg/mi)	Fuel 63 Specific Reactivity	Fuel 63 Log Total Reactivity	Fuel 63 Log Specific Reactivity	Fuel 64 Total Reactivity O_3 (mg/mi)	Fuel 64 Specific Reactivity	Fuel 64 Log Total Reactivity	Fuel 64 Log Specific Reactivity
7		19074.88	1.72			27754.89	1.61		
		19707.16	1.74			24402.44	1.63		
7	*Mean*	19391.02	1.73	*4.29*	*0.24*	26078.66	1.62	*4.42*	*0.21*
8		14637.40	1.36			25633.44	1.37		
		13567.74	1.33			27628.16	1.40		
8	*Mean*	14102.57	1.34	*4.15*	*0.13*	26630.80	1.38	*4.43*	*0.14*
9		3061.17	1.28			6595.22	1.45		
		3392.42	1.36			6974.17	1.37		
9	*Mean*	3226.80	1.32	*3.51*	*0.12*	6784.69	1.41	*3.83*	*0.15*
10		11190.77	1.69			33313.56	1.65		
		14347.75	1.78			29459.05	1.74		
		16287.71	1.45						
10	*Mean*	13942.08	1.64	*4.14*	*0.22*	31386.30	1.69	*4.50*	*0.23*
11		3217.83	1.38			7084.59	1.28		
		2879.87	1.54			9709.03	1.26		
11	*Mean*	3048.85	1.46	*3.48*	*0.16*	8396.81	1.27	*3.92*	*0.10*
13		4745.35	1.13			4335.13	1.25		
		2299.48	1.21			4959.99	1.18		
		2662.13	1.17						
13	*Mean*	3235.65	1.17	*3.51*	*0.07*	4647.56	1.21	*3.67*	*0.08*
	Mean	9490	1.44	*3.85*	*0.156*	17300	1.43	*4.13*	*0.152*
	sd	7200	0.212	*0.383*	*0.0638*	11900	0.191	*0.360*	*0.0572*
	sdm	2940	0.0867	*0.156*	*0.0261*	4870	0.0778	*0.147*	*0.0233*
t test		F-64-F-63	F-64-F-63	F-64-F-63	F-64-F-63				
p value		0.024	0.77	0.0022	0.81				